Simulation of Automotive Radar Point Clouds in Standardized Frameworks

TECHNISCHE UNIVERSITÄT MÜNCHEN
Fachgebiet Höchstfrequenztechnik
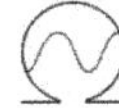

Simulation of Automotive Radar Point Clouds in Standardized Frameworks

Thomas Eder

Vollständiger Abdruck der von der Fakultät für Elektrotechnik und Informationstechnik der Technischen Universität München zur Erlangung des akademischen Grades eines

Doktors der Ingenieurwissenschaften (Dr. Ing.)

genehmigten Dissertation.

Vorsitzende: Prof. Dr. Eva Weig
Prüfer der Dissertation:
1. Prof. Dr.-Ing. Erwin Biebl
2. Prof. Dr.-Ing. Walter Stechele

Die Dissertation wurde am 20.05.2021 bei der Technischen Universität München eingereicht und durch die Fakultät für Elektrotechnik und Informationstechnik am 30.08.2021 angenommen.

Bibliografische Information der Deutschen Nationalbibliothek
Die Deutsche Nationalbibliothek verzeichnet diese Publikation in der Deutschen Nationalbibliografie; detaillierte bibliographische Daten sind im Internet über http://dnb.d-nb.de abrufbar.
1. Aufl. - Göttingen: Cuvillier, 2021
Zugl.: (TU) München, Univ., Diss., 2021

Nonnenstieg 8, 37075 Göttingen
Telefon: 0551-54724-0
Telefax: 0551-54724-21
www.cuvillier.de

1. Auflage, 2021
Gedruckt auf umweltfreundlichem, säurefreiem Papier aus nachhaltiger Forstwirtschaft.

ISBN 978-3-7369-7536-1
eISBN 978-3-7369-6536-2

Essentially, all models are wrong, but some are useful.

George E. P. Box

Abstract

Driver assistance systems have evolved rapidly in recent years and are an indispensable feature of many of today's vehicle models. The increasing trend of industry and research is certainly towards autonomous driving. The complexity of the system and the complexity of the test cases can no longer be covered by real driving tests in an economically sustainable way. For this reason, researchers and industry experts are trying to replace actual test drives with simulation-based tests designed to be as close to reality as possible. The simulation of the vehicle's environmental sensors, the so-called sensor simulation, is therefore crucial. Automobile manufacturers are increasingly focusing on a standardized architecture with a high level of abstraction. In order to simulate the sensors, such as radar sensors, most realistically on a point cloud level, data-based methods are used in many cases. In general, and specifically in case of radar sensors, there are still challenges to be faced. Therefore, four research questions are addressed:
Is a certain physical model of a frequency modulated continuous wave radar sensor accurate enough to generate synthetic training data for data-based models? This question is answered by means of a thorough and detailed investigation of the underlying method and the associated approximations. Among the results are a detailed error analyses based on an exact formula for the beam expansion derived herein.
Which statistical approaches are suitable for the simulation of entire radar point clouds and how shall their learning capacities be evaluated? Generative neural networks and kernel density estimation methods for point cloud simulation are developed. Based on exemplary metrics, the methods are compared objectively. Finally, a subjective evaluation from the developer's point of view regarding the genericity of the methods is given.
Furthermore, a generic and user-friendly hybrid approach is developed to circumvent the disadvantages of purely statistical techniques. Based on a simple ray-tracing based approach, a radar detection model is developed that takes simplified physical laws into account and may be fitted to a specific sensor without the need of large amounts of data. Finally, numerous opportunities for enhancement are discussed.
The problem of evaluation and validation of radar point cloud models is discussed and a consistency criteria is proposed and discussed. A simple example highlights current challenges and is followed by the presentation of a criterion for consistency of validation methods, which is tested by means of statistical tests presented herein for the first time.

Acknowledgement

The more than three years, during which this doctoral thesis was written, were marked by many pleasant moments. Moments of liberty, but also moments of considerable workload. Unfortunately, this time was also affected by strokes of fate. But eventually, all things merge into one. However, I owe a great debt of thanks to many kind and helpful people. Especially, to my doctoral advisor, Prof. Dr.-Ing. Erwin Biebl. Thank you very much for the great cooperation, for the integration at your chair, for your understanding, and patience. But also for supervising the thesis over all these years and for your constant support.

I would also like to thank Carlo van Driesten, without whom the whole project would not have come into being. Thank you for making it all possible in the first place. Thank you for your understanding, the exchange and your talent to always make the best use of the financial resources and to remove all impediments.

I would like to thank all my colleagues from the Technical University of Munich for wonderful memories in Kleinheubach and the numerous telephone conferences. Especially to Onur Kepenek and Christian Buchberger for great conversations, interesting discussions and organizational support during the lectures.

This work was done in cooperation with the BMW AG. I was privileged to spend the majority of my doctoral period there. Such an opportunity is by no means self-evident for me, which is why I would like to thank the entire company and especially the program management of the doctoral program. In addition, I am still indebted to my students and many other people, all of whom I cannot possibly name here. In particular, I would like to thank Lavinia Graff, Valentin Protschky, Alexander Prinz, Egon Ye, Alexander Frickenstein and many, many others!

I also owe a lot to my former teachers Robert Liesaus (né Hering), Michael Schmid and Martin Schwingenheuer. Many thanks, I probably would not have studied mathematics and physics without you.

Eventually, however, my utmost gratitude belongs to my family and my beloved ones, who have always been supportive and motivating. Especially to my deceased father and my caring stepmother. My beloved wife, Susie Marie Eder, has contributed significantly to the success of this endeavor. Thank you for proofreading. Thank you for your support all along the way. And thank you for your understanding and perseverance.

Contents

1 Autonomous driving and simulational challenges **1**
1.1 Safety validation and simulative test drives 1
1.2 Principles of automotive radar sensors 3
1.2.1 Static targets and their radar signal 4
1.2.2 The Doppler effect and its influence 6
1.2.3 Angle estimation with multiple receivers 9
1.2.4 Radar signal processing and point clouds 10
1.3 Modeling and standardized simulation frameworks 14

2 State of research in automotive radar modeling **17**
2.1 Differentiation of various modeling levels 17
2.2 Ray-tracing in environments of high-fidelity 20
2.3 Models executable in standardized environments 21
2.3.1 Simplified physical models 22
2.3.2 Statistical sensor models 24
2.4 Validation and verification of sensor models 25

3 Derivation of research questions, hypotheses and objectives **29**
3.1 Inaccuracies of current ray cone tracing based models 29
3.2 Stochastic radar models based on deep generative networks 30
3.3 Hybrid multipurpose approaches for radar sensor models 31
3.4 Deficiencies of current validation criteria 32

4 Modeling challenges related to ray cone tracing **33**
4.1 The caustic distance and the angular beam expansion 33
4.2 Estimating current errors in case of multiple reflections 37
4.3 Consequences and lower bounds for the number of rays 40

5 Approaches to statistical radar point cloud simulation **43**
5.1 Statistical formulation of radar sensor modeling 43
5.2 Kernel density estimation and radar point clouds 46
5.2.1 Principles of kernel density estimation 46

5.2.2 Multivariate application to radar point clouds 49
5.3 Deep generative networks as sensor models 51
5.3.1 Variational Autoencoder 52
5.3.2 Generative Adversarial Networks 54
5.4 Comparison of learning capacities and its consequences 56

6 A hybrid modeling approach for radar point clouds 59
6.1 Tracing and catching rays as the baseline 59
6.2 Improvements to the ray casting approach 64
6.2.1 Characteristic scattering centers 64
6.2.2 The Doppler effect of wheels and tires 66
6.3 Capabilities for data-based optimization 72
6.3.1 Adapting the distance-dependent measurement likelihood 73
6.3.2 The spatial measurement deviation of scattering centers 74
6.4 Bottom line on the hybrid modeling approach 75

7 Validation based on statistical hypothesis testing 77
7.1 Consistency of validation criterion 77
7.2 On the Kolmogorov–Smirnov test 79
7.3 Applications to radar sensor models 83
7.3.1 Scenario based application 83
7.3.2 Position based application 85
7.4 Retrospective and future validation challenges 87

8 Conclusion and prospective challenges 89
8.1 Recap of the radar point cloud simulation 89
8.2 Lessons learned and future recommendations 91

Nomenclatur 93

References 99

Index 109

Chapter 1

Autonomous driving and simulational challenges

For more than one hundred years the automotive industry has been a warrant for technological innovation, sustainable employment and prosperity. With the introduction of seat belts and airbags vehicles got increasingly safer. The development of an electronic fuel injection system lead to severely diminished emissions due to lower fuel consumption. Moreover, and despite the rising technological complexity prices dropped considerably with regards to the Consumer Price Index. As a summary it may be stated that motor vehicles got cleaner, safer, and more affordable. The automobile as a symbol of freedom, indepence, and prosperity lead to a continually increasing demand. In conjunction with additional political measures the automotive sector has become the leading industry in many industrialized nations like the United States, China, and Germany, to name just a few.

In spite of all this, time does not stand still and so car manufactures are facing great upheavals nowadays. For a long time period California has been a sole pioneer for tightening emission regulations and thus increased the demand for electric vehicles. While senior industry experts kept saying lithium-ion technology is still several years away Tesla introduced the first all-electric car based upon lithium-ion battery cells in 2008. Even though new competitors are struggling with numerous difficulties till date their way of working as well as their influence to current and future development projects in the automotive sector is already considerable. Economy and society are on the brink of another new technological revolution: autonomous driving [1].

1.1 Safety validation and simulative test drives

The almost one hundred year old idea of self-driving cars might solve several intractable problems related to everyone's daily mobility. These include, among others, traffic congestions, availability of public transport, medical or age-related mobility constraints as well as the often quoted number of road fatalities. The particular interest of many companies in this technology, however, is especially based on associated disruptive business models [2].

Fig. 1.1. Schematic, exemplary view of the various sensors around a vehicle with a driver assistence system. The picture is taken from a highway scene within a virtual environment. The colored sensor cones show each sensors approximate field of view. The different colors indicate different types of sensors, e.g., radar, lidar, and camera sensors (see also [3]).

However, the almost 25-year-long development of adaptive cruise control (ACC) has demonstrated that innovation development up to the stage of production readiness requires a certain amount of effort. On the one hand, there was uncertainty about the customer features and the technology to be used (pulse-width modulation or frequency modulation), and on the other hand, there was a lack of suitable suppliers. From today's point of view, it is not surprising that numerous product releases failed due to the extensive efforts required for integration and deployment, as well as the difficulty in communicating the resulting benefits to the customers. Nevertheless, adaptive cruise control has matured to become a mass product these days, paving the way for the further development of advanced driver assistance systems.

The development of autonomous driving and modern advanced driver assistance systems always focuses on controllability by the driver and constitutes the basis for safety validation of individual functions. This applies in particular to conditional automation, the so-called Level 3 of driving automation according to the SAE J3016 standard. Here, all aspects of the driving task are performed by an autonomous driving function, whereby the driver is always expected to be able to intervene on demand in case of doubt. Typically, the driver is expected to be ready to take over the driving task within a few seconds. This strategy leads to enormous demands on the view of environmental sensors. However, particularly due to the fact that even the simplest tasks are covered by the autonomous driving function, even the very simplest but rarely occurring events must be safe guarded prior to a vehicle's release. In case of events that occur only after an average of 50,000 kilometers or more (see 1.2), such as an emergency break event, several million test kilometers have to be performed beforehand. Additional requirements for a particular test concept, such as weather conditions or configuration options, may again drastically increase the number of test kilometers to be covered [4, 5].

When deriving the number of test kilometers, a Poisson distribution is usually assumed. For example, if a certain event occurs n times on average for a distance length s_0, then

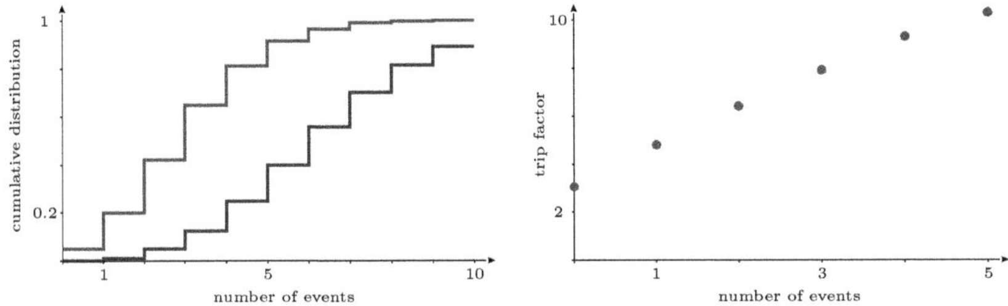

Fig. 1.2. Effects on the number of test kilometers assuming an underlying Poisson distribution. On the left, the cumulative probability for the number of occurrences, if $\lambda = 3$ (blue) or $\lambda = 6.3$ (red) events are to be expected during the test. On the right, the stretch factor for the test drive in case of a given number of events (see also [4, p. 1175 ff.]) for further details.

it is expected that in case of a trip length s the event occurs $s/s_0 \times n$ times. However, this does not guarantee that the event will occur that often. Instead, this value forms the expectation value λ of the Poisson distribution. For instance, if an event occurs on average only once at a distance s_0, then at the three-fold distance, i.e., $\lambda = 3$, the event does not occur at all in 5% of the cases (see 1.2). This is referred to as significance level and is usually tolerated up to a probability of 5%. Therefore, for a given event, at least three times the distance must be completed without an accident. Alternatively, 6.3 times with a maximum of one accident.

Such a safety validation strategy rapidly leads to a coverage of up to multiple hundred million test kilometers and exceeds the technical, personnel and financial capabilities of today's companies. Even if this effort was ventured for an initial system under limited circumstances, one has to consider that the test would have to be performed again with at least one third of the initial effort after each modification of the system, which is obviously not economically justifiable [4]. Therefore, the prevailing opinion among automotive manufacturers, suppliers and also within the research community tends towards a decisive role for simulation-based safety validation in the future [6].

1.2 Principles of automotive radar sensors

The first radar systems originated in the field of military and aviation technology. For about two decades, these sensors have also been increasingly integrated into automobiles. Today, upper-class vehicles already include numerous radar sensors that provide data for various driver assistance functions. In general, however, vehicle manufacturers only specify certain sensor requirements and subsequently either purchase these sensors or develop them in collaboration with suppliers.

Regardless of whether the design of models is performed by the sensor manufacturer himself or by automotive or simulation manufacturers themselves, it is crucial to understand the fundamental principles of radar sensor technology and to be familiar with the definition of the interfaces. Essentially, to determine the distance of an object, a radar sensor usually measures the time of flight t taken by an electromagnetic wave to travel back and forth between sensor and target. Hence the distance R is given by

$$R = \frac{c_0}{2} t, \tag{1.1}$$

whereby c_0 is the speed of light. For automotive applications, however, so called pulsed radar systems, which measure the runtime directly, have numerous disadvantages. Thus, targets can only be distinguished from each other if the distance between them is sufficiently large, i.e., depending on the pulse duration t_p the distance must be at least

$$\Delta R \geqslant \frac{c_0}{2} t_p. \tag{1.2}$$

Due to legal restrictions and the fact that the -3 dB bandwidth is inversely proportional to the pulse duration, t_p cannot be chosen arbitrarily small. Therefore, the frequency modulated continuous wave (FMCW) measurement principle is widely accepted and currently de facto standard for automotive purposes. Therefore, a short introduction to the essential principles of modern automotive radar sensors will be given in the following. The subsection focuses primarily on signal processing and concludes with a general description of the radar point cloud interface.

1.2.1 Static targets and their radar signal

Frequency modulation is characterized by the transmission of a continuous wave, with periodically increasing or decreasing frequency. The following explanation deals with the common case of sawtooth frequency modulation, i.e., the time dependent transmit frequency f_{Tx} is given by

$$f_{\mathrm{Tx}}(t) = f_{\mathrm{c}} + \frac{B}{T}\left(t - \left\lfloor \frac{t}{T} \right\rfloor T\right). \tag{1.3}$$

Each sawtooth, also called chirp, passes through the frequency range from f_{c} to $f_{\mathrm{c}} + B$. Thereby, f_{c} indicates the carrier frequency, B the bandwidth and T the chirp or sweep time. In case of a reflective target at a distance R, the signals time of flight is

$$\tau_0 = \frac{2R}{c_0}. \tag{1.4}$$

This results in a time dependent receive frequency

$$\begin{aligned} f_{\mathrm{Rx}}(t) &= f_{\mathrm{Tx}}(t - \tau_0) \\ &= f_{\mathrm{c}} + \frac{B}{T}\left(t - \left\lfloor \frac{t}{T} \right\rfloor T - \tau_0\right). \end{aligned} \tag{1.5}$$

Now, if the difference frequency Δf between the transmit and receive frequency is measured, the distance R to the target can be determined, i.e.,

$$R = \frac{c_0 T \Delta f}{2B}. \tag{1.6}$$

To be more specific about digital signal processing and the frequency measurement principles, it is necessary to examine the actual transmit and receive signal more closely. The signal of an unmodulated wave at frequency f is given at the transmitting antenna as

$$\begin{aligned} u_{\mathrm{Tx}}(t) &= u_0 \cos\left(\varphi_{\mathrm{Tx}}(t)\right) \\ &= u_0 \cos\left(2\pi f t\right). \end{aligned} \tag{1.7}$$

Thereby, the factor u_0 denotes the amplitude and φ_{Tx} the phase of the transmit signal. To derive a more profound interrelation between frequency and phase of a frequency modulated signal the term instantaneous frequency must first be explained. Since frequency is the number of revolutions (or simply repetitions) per time, the formula can be expressed in terms of the phase angle as well. Therefore, the average frequency $\bar{f}_h$ in a small interval $[t, t+h]$ of length h is given as

$$\bar{f}_h(t) = \frac{1}{2\pi} \frac{\varphi(t+h) - \varphi(t)}{h}. \tag{1.8}$$

Now, the limit $h \to 0$ defines the instantaneous frequency. Thus, it can be concluded that frequency is the time derivative of the phase, i.e.,

$$f(t) = \frac{1}{2\pi} \frac{\mathrm{d}}{\mathrm{d}t} \varphi(t). \tag{1.9}$$

Conversely, the phase can be determined by integrating the time dependent frequency. Hence the phase is given by

$$\varphi_{\mathrm{Tx}}(t) = 2\pi \int_0^t f_{\mathrm{Tx}}(u) \mathrm{d}u. \tag{1.10}$$

Provided that all chirps start in phase, it suffices to simply integrate from 0 to $(t - \lfloor t/T \rfloor T)$ to determine $\varphi_{\mathrm{Tx}}(t)$. An analogous formula applies to the phase of the receive signal φ_{Rx} at $t \geqslant \tau_0$, i.e.

$$\begin{aligned} \varphi_{\mathrm{Rx}}(t) &= 2\pi \int_{\tau_0}^t f_{\mathrm{Rx}}(u) \mathrm{d}u \\ &= 2\pi \int_0^{t-\tau_0} f_{\mathrm{Rx}}(u+\tau_0) \mathrm{d}u \\ &= 2\pi \int_0^{t-\tau_0} f_{\mathrm{Tx}}(u) \mathrm{d}u \\ &= \varphi_{\mathrm{Tx}}(t-\tau_0). \end{aligned} \tag{1.11}$$

The intermediate frequency signal u_Δ of u_{Tx} and u_{Rx} is a nearly continuous wave with frequency Δf. Technically, the differential signal is generated by means of a multiplicative mixer and a subsequent low-pass filter. The mixer multiplies both signals with each other. According to the trigonometric product-sum identity, the resulting signal u_* has the form

$$\begin{aligned} u_*(t) &= u_{\mathrm{Tx}}(t) u_{\mathrm{Rx}}(t) \\ &= u\left(\cos\left(\varphi_{\mathrm{Tx}}(t) + \varphi_{\mathrm{Rx}}(t)\right) + \cos\left(\varphi_{\mathrm{Tx}}(t) - \varphi_{\mathrm{Rx}}(t)\right)\right). \end{aligned} \tag{1.12}$$

In general, since the receive signal passes through an amplifier before, there is an additional constant $c > 0$ such that $u = u_0 u_1 c$. In the context of signal processing, the exact value of the amplitude can usually be neglected anyway. Now, the low-pass filter removes all high-frequency components, hence only the difference frequency remains present. Consequently, the intermediate frequency signal is given by

$$u_\Delta(t) = u \cos\left(\varphi_{\mathrm{Tx}}(t) - \varphi_{\mathrm{Rx}}(t)\right). \tag{1.13}$$

According to equation (1.10) and (1.11) one can derive a detailed expression for the intermediate frequency signal. However, since this leads to an extremely elongated expression, the signal is usually only given for the range of constant intermediate frequency. This range extends from τ_0 to $T - \tau_0$ and accounts for considerably more than 90% of the signal. For example, a signal with $T = 20\mu s$ contains 99% constant frequency components for a target at a distance of 15m. For the previously mentioned range the temporal signal is given by

$$\begin{aligned} u_\Delta(t) &= u \cos\left(2\pi\left(\frac{B}{T}\tau_0\left(t - \left\lfloor\frac{t}{T}\right\rfloor T\right) + f_c\tau_0 - \frac{B}{2T}\tau_0^2\right)\right) \\ &\approx u \cos\left(2\pi\left(\frac{B}{T}\tau_0\left(t - \left\lfloor\frac{t}{T}\right\rfloor T\right) + f_c\tau_0\right)\right). \end{aligned} \tag{1.14}$$

Since a window function is usually applied to the intermediate frequency signal anyway, it is sufficient for almost all applications and models to consider a signal of constant intermediate frequency. For consideration of more complex effects due to the time ranges $[0, \tau_0]$ and $[T - \tau_0]$, the phase must be calculated for the full range. In addition to a different phase, and depending on the window function, there is also a time-dependent amplitude. The following neglects these effects and proceeds with the formula given in equation (1.13).

1.2.2 The Doppler effect and its influence

For the derivation of the frequency modulated continuous wave radar signal only static targets have been considered so far. However, if the target moves the frequency of the received signal is shifted due to the Doppler effect. For the receive frequency f_{Rx} of a moving target at the radial speed v_{r}, the following applies

$$f_{\mathrm{Rx}} = \frac{c_0 - v_{\mathrm{r}}}{c_0 + v_{\mathrm{r}}} f_{\mathrm{Tx}}. \tag{1.15}$$

Since an exact relativistic consideration is usually not necessary at low speeds, the above formula is often simplified by a first-order taylor approximation. Therefore, equation (1.15) can be written as

$$f_{\mathrm{Rx}} \approx \left(1 - \frac{2v_{\mathrm{r}}}{c_0}\right) f_{\mathrm{Tx}}. \tag{1.16}$$

As is the case with a primary radar, the above formula given in equation (1.15) considers two relativistic Doppler shifts. The first shift occurs during the reception at the target.

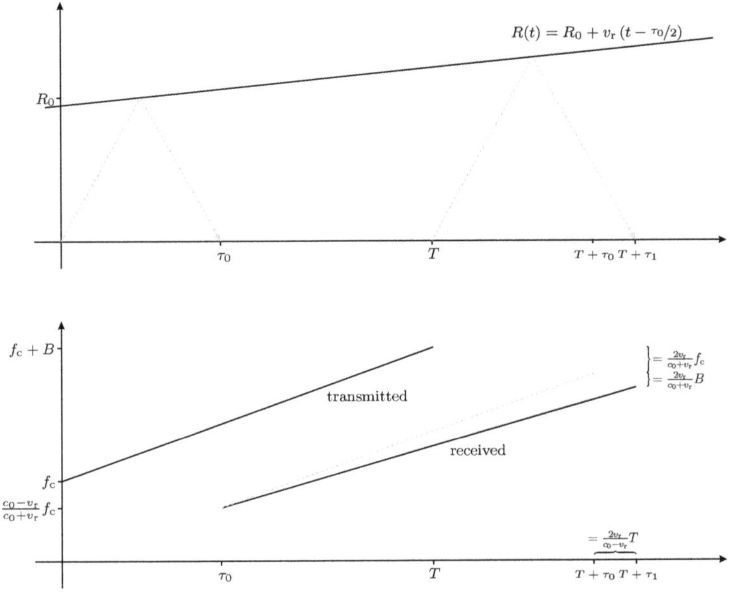

Fig. 1.3. Illustration of the relativistic Doppler effect and its influence on a single chirp. During the transmission of a chirp, the target vehicle continuously moves and as a result, a frequency shift occurs. Since the chirp duration is also subject to a Doppler shift, the graphs of transmitted and received frequencies are not perfectly parallel. However, these nuances are generally negligible.

The second one when the signal reaches the transmitter after reflection. For example, if the target moves radially away from the sensor, $v_r > 0$ applies and thus $f_{Rx} < f_{Tx}$. In order to finally reproduce such a frequency shift within the phase of the time signal u_Δ, an exact relativistic consideration of the received phase is necessary. In numerous publications and contributions one can find simpler derivations, which only consider a time-dependent runtime τ in equation (1.11). However, this reproduces the Doppler shift only approximately and neglects certain effects due to frequency modulation (see the following remark).

In order to determine the phase of the reflected signal, in case of a single dynamic target, a twofold coordinate transformation must be performed. For this purpose, the transmitted spacetime-signal (for simplicity limited to only one chirp)

$$u_{Tx}(t, x) = u_0 \cos\left(2\pi\left(f_c\left(t - \frac{x}{c_0}\right) + \frac{B}{2T}\left(t - \frac{x}{c_0}\right)^2\right)\right) \tag{1.17}$$

is transformed into the reference frame of the target. According to the Lorentz transformation, the coordinates (t, x) are given in the reference frame of the target as

$$\begin{aligned} x' &= \gamma\left(x + v_{\mathrm{r}} t\right) \\ t' &= \gamma\left(t + \frac{v_{\mathrm{r}} x}{c_0^2}\right), \end{aligned} \tag{1.18}$$

whereby $\gamma = \left(1 - {}^{v_{\mathrm{r}}^2}\!/\!{}_{c_0^2}\right)^{-1/2}$ is the Lorentz factor [7]. Plugging (t', x') into equation (1.17) one gets the transmit signal from the perspective of the target vehicle. Now, a reflection is equivalent to the instantaneous emission of the signal received from the sensor. Due to the principle of relativity, instead of a moving target and a fixed sensor, one can just as well assume a stationary target and a moving sensor. Therefore, the reflected signal in the sensor reference frame is obtained by another Lorentz transformation. Just with exchanged roles of (t', x') and (t, x). If one considers the additional time of flight τ_0, this results in the following overall transformation

$$\left(t - \frac{x}{c_0}\right) \rightsquigarrow \frac{c_0 - v_{\mathrm{r}}}{c_0 + v_{\mathrm{r}}}\left(t - \frac{x}{c_0}\right) \rightsquigarrow \frac{c_0 - v_{\mathrm{r}}}{c_0 + v_{\mathrm{r}}}\left(t - \tau_0 - \frac{x}{c_0}\right) \tag{1.19}$$

for the transition from u_{Tx} to u_{Rx}. The phase of the reflected signal in the reference frame of the sensor at $x = 0$, i.e., the phase of the received signal, is therefore given as

$$\varphi_{\mathrm{Rx}}(t) = 2\pi \frac{c_0 - v_{\mathrm{r}}}{c_0 + v_{\mathrm{r}}}\left(f_{\mathrm{c}}\left(t - \tau_0\right) + \frac{c_0 - v_{\mathrm{r}}}{c_0 + v_{\mathrm{r}}} \frac{B}{2T}\left(t - \tau_0\right)^2\right). \tag{1.20}$$

REMARK. Deviding equation (1.20) by 2π and performing a time derivative, one obtains the instantaneous receive frequency

$$f_{\mathrm{Rx}}(t) = \frac{c_0 - v_{\mathrm{r}}}{c_0 + v_{\mathrm{r}}}\left(f_{\mathrm{c}} + \frac{c_0 - v_{\mathrm{r}}}{c_0 + v_{\mathrm{r}}} \frac{B}{T}\left(t - \tau_0\right)\right)$$

Now, one clearly sees that the Doppler shift affects the current transmit frequency on the one hand, but also the chirp frequency ${}^{1}\!/\!{}_{T}$ on the other. Consequently, the duration of the chirp also changes (see Fig. 1.3). The latter two facts are usually neglected.

To obtain a concise, closed form for the intermediate frequency signal u_Δ, again several first-order Taylor approximations are necessary. An extensive calculation then yields

$$u_\Delta(t) \approx u \cos\left(2\pi\left(\frac{B}{T}\tau_0 t + f_{\mathrm{c}}\tau_0 - \frac{B}{2T}\tau_0^2 + \frac{2v_{\mathrm{r}}}{c_0} f_{\mathrm{c}}\left(t - \tau_0\right) + \frac{4v_{\mathrm{r}}}{c_0} \frac{B}{2T}\left(t - \tau_0\right)^2\right)\right) \tag{1.21}$$

for a single chirp. In the limiting case $v_{\mathrm{r}} \to 0$, one obtains the already known formula for the static case presented in equation (1.14). In practice, an additional approximation is usually performed, and hence one usually considers

$$u_\Delta(t) \approx u \cos\left(2\pi\left(\left(\frac{B}{T}\tau_0 + \frac{2v_{\mathrm{r}}}{c_0} f_{\mathrm{c}}\right) t + f_{\mathrm{c}}\tau_0\right)\right) \tag{1.22}$$

only. For example, this approach is also used in [8, p. 49 ff.] but derived differently (see the following remark). The intermediate frequency signal u_Δ contains not only a time of flight dependent part, but also an additional frequency shift due to the Doppler effect[1]. Therefore, the objective of radar signal processing is to separate the time-of-flight component from the Doppler component as precisely as possible in order to enable simultaneous measurement of distance and velocity. Subsequent to the basics of angle estimation, the signal processing of modern radar sensors is explained taking this derivation into account.

REMARK. The formula given in equation (1.22) is usually derived by simpler methods than a Lorentz transformation, see [8, p. 49 ff.] for example. However, from the physical point of view it is not entirely correct. Nevertheless, since it is common practice, a possible procedure is briefly presented here.
Suppose the target moves with constant radial speed, then the distance R changes linearly and so does the runtime τ_0. Hence the time-dependent runtime is given by

$$\tau(t) = \tau_0 + \frac{2v_\mathrm{r}}{c_0}t. \tag{1.23}$$

If one proceeds analogously to the static case as described in equation (1.11) and uses the above relation during integration, one obtains the following approximation

$$\begin{aligned} \varphi_\mathrm{Rx}(t) &= 2\pi \int_{\tau(t)}^{t} f_\mathrm{Rx}(u)\mathrm{d}u \\ &\approx 2\pi \left(\left(1 - \frac{2v_\mathrm{r}}{c_0}\right)\frac{B}{2T}t^2 + \left(\left(1 - \frac{2v_\mathrm{r}}{c_0}\right) f_\mathrm{c} - \frac{B}{T}\tau_0\right) t - f_\mathrm{c}\tau_0 \right), \end{aligned} \tag{1.24}$$

if terms of order v_r^2/c_0^2, v_r^3/c_0^3 and τ_0^2 are neglected. After calculating the phase difference as indicated in (1.13) and neglecting terms of order t^2, one obtains the the same intermediate frequency signal for the dynamic case as presented in equation (1.22).

1.2.3 Angle estimation with multiple receivers

Measuring the speed and distance of a target vehicle is still not sufficient for an environmental perception. In addition, an angular measurement in azimuthal direction is also necessary. In order to evaluate time-of-flight dependent differences and thus determine the direction of incidence, radar systems require at least two receiving antennas. Assuming a reflective target is located in polar coordinates at (R_0, θ) with $\theta > 0$, the reflected signal will reach the left receiving antenna first. Although the time difference is minimal due to the speed of light, the phase of the received signal changes (see Fig. 1.4). The ratio of the phase difference $\Delta\varphi$ and 2π matches the ratio of the additional path length ΔR and the wavelength λ, i.e.,

$$\frac{\Delta\varphi}{2\pi} = \frac{\Delta R}{\lambda}. \tag{1.25}$$

[1] Using equation (1.22), the non-parallelity of transmitted and received chirps (see Fig. 1.3) is neglected.

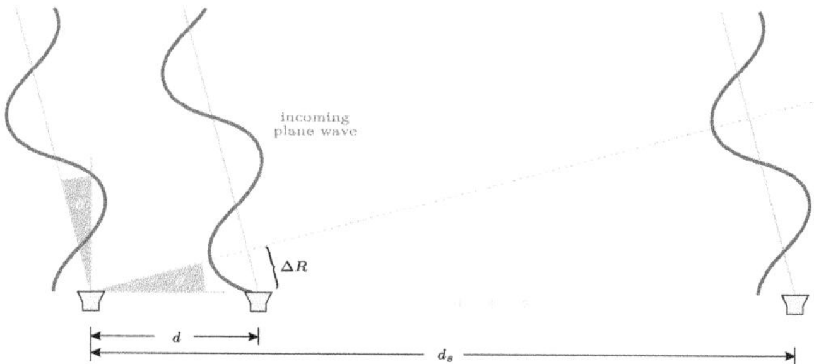

Fig. 1.4. Illustration of the angle measurement principle. The antenna distance d_s of the fictitious receiving element on the right side is chosen in a way that a phase difference of 2π, respectively λ arises compared to the first receiving antenna. This argument is used within the signal processing subsection.

The additional path depends in good approximation (plane wavefront) only on the angle of incidence θ and the spacing between the antennas d. Thereby

$$\Delta R = d \sin\theta \tag{1.26}$$

applies. Since the phase difference must be in the range from $-\pi$ to π, the antenna spacing is also directly related to the unique angular range of the sensor. For a common antenna spacing of $\lambda/2$, a maximum range of $-90°$ to $90°$ is obtained due to

$$\alpha = \sin^{-1}\left(\frac{\lambda}{d}\frac{\Delta\varphi}{2\pi}\right). \tag{1.27}$$

Due to the non-linearity, however, a significantly larger angular error results at the boundaries of the range of uniqueness even in case of slightest phase measurement errors.
This simple principle is already sufficient for determining the angle of a single target. In practice, though, often more than two antennas are used. The exact way in which the phase difference is measured is discussed in the following section.

1.2.4 Radar signal processing and point clouds

The purpose of signal processing is to determine the distance, the velocity and the incidence angle of a target. Essentially, the signal processing is based on the frequency analysis by means of Fourier transformations, which is explained step by step in the following. The subsequent detection is generated with a constant false alarm rate algorithm, which will be briefly discussed afterwards.

Frequency analysis with Fourier transformations. If the signal from equation (1.21), resp. (1.22) is received by the receiving antenna, the analog signal is sampled at a frequency f_AD using an analog-to-digital converter. Therefore, a chirp of length T

yields exactly $d = T \cdot f_{\text{AD}}$ sampling points, which corresponds to a d-dimensional vector (f_{AD} is chosen such that d is an integer). Using the discrete Fourier transform (usually implemented in terms of a fast Fourier transform), the amplitude signal is transformed from the time domain to the frequency domain, where

$$\begin{aligned} &\mathfrak{F}^{(d)} : \mathbb{R}^d \to \mathbb{C}^d \\ &(a_0, a_2, \ldots, a_{d-1}) \mapsto (\hat{a}_0, \hat{a}_2, \ldots, \hat{a}_{d-1}) \quad \text{whereby} \quad \hat{a}_k = \sum_{j=0}^{d-1} a_j \mathrm{e}^{-2\pi \mathrm{i} \cdot \frac{jk}{d}}. \end{aligned} \tag{1.28}$$

Just the same way as the values $(a_0, a_2, \ldots, a_{d-1})$ are related to the sampling timestamps

$$(0, 1, \ldots, d-1) \cdot \frac{1}{f_{\text{AD}}}, \tag{1.29}$$

the transformed values $(\hat{a}_0, \hat{a}_2, \ldots, \hat{a}_{d-1})$ correspond to the frequencies

$$(0, 1, \ldots, d-1) \cdot f_{\text{AD}}. \tag{1.30}$$

If, for example, the value $\hat{a}_i$ corresponds to a strong peak in the frequency spectrum, then one can determine the distance of the object with the equation (1.6) and the corresponding frequency $i \cdot f_{\text{AD}}$. However, the following relationship must be taken into account: The discrete Fourier transform exhibits a certain symmetry, thus for real samples

$$\begin{aligned} \hat{a}_{d-k} &= \sum_{j=0}^{d-1} a_j \mathrm{e}^{-2\pi \mathrm{i} \cdot \frac{jd}{d}} \mathrm{e}^{2\pi \mathrm{i} \cdot \frac{jk}{d}} \\ &= \sum_{j=0}^{d-1} \overline{a_j \mathrm{e}^{-2\pi \mathrm{i} \cdot \frac{jk}{d}}} \\ &= \overline{\hat{a}_k} \end{aligned} \tag{1.31}$$

applies. Consequently, there are at most $d/2$ independent frequencies (if d is even, then $d/2$ is the Nyquist frequency). Together with the Nyquist-Shannon sampling theorem, it follows that only $\lceil (d-1)/2 \rceil$ values are relevant, and therefore only the frequencies

$$\left(0, 1, \ldots, \left\lceil \frac{d-1}{2} \right\rceil \right) \cdot f_{\text{AD}} \tag{1.32}$$

must be considered for the range evaluation. In summary, it can be stated that the distance is essentially determined by means of the frequency analysis of a single chirp. The error contained therein caused by the Doppler shift can be neglected, since in practical applications the error is smaller than the distance of the frequency bins.

To determine not only the distance of an object but also its velocity, several chirps are required. If, for example, n_{C} chirps are emitted coherently during a measurement cycle, the j-th sample point of each chirp can be interpreted as a sample of a continuous wave. Hence, the frequency corresponds directly to the frequency shift induced by the Doppler effect[2]. Considering equation (1.16), one finds that the intermediate frequency fulfills

$$\begin{aligned} \Delta f &= \frac{2v_\mathrm{r}}{c_0}\frac{1}{T} \\ \Leftrightarrow \qquad v_\mathrm{r} &= \frac{\lambda}{2}\Delta f. \end{aligned} \tag{1.33}$$

Thus, if one performs a Fourier transform in the second dimension along all chirps, one can convert the corresponding frequency components into velocities using equation (1.33). A similar procedure is used for the determination of the angle. As already indicated in subsection 1.2.3, the angle can be determined by means of the delay-induced phase differences at several receiving antennas. Considering the difference across multiple antennas (see Fig. 1.4), a spatial frequency f_s, whose period is denoted by d_s, is obtained by sampling at the same time at all receiving antennas. Since at a fictitious antenna spacing d_s the phase difference corresponds to 2π and thus to a wavelength λ, analogous to equation (1.26) the following holds

$$\begin{aligned} \lambda &= d_\mathrm{s} \sin\theta \\ \Leftrightarrow \qquad \theta &= \sin^{-1}\left(\lambda f_\mathrm{s}\right). \end{aligned} \tag{1.34}$$

Consequently, a third Fourier transform over the different antenna elements can be used to determine the angle of incidence θ based on the phase difference. The sampling values are often denoted as

$$a_{i,j}^{(k)} \quad \text{for } 1 \leqslant i \leqslant \lceil (d-1)/2 \rceil,\ 1 \leqslant j \leqslant n_\mathrm{C},\ \text{and } 1 \leqslant k \leqslant n_\mathrm{Rx} \tag{1.35}$$

in the case of n_C chirps and n_Rx receiving antennas. After threefold Fourier transformation one obtains the three-dimensional tensor $\left\{A_{i,j}^{(k)}\right\}$, which is often also called radar cube (with the dimensions range, velocity, and angle). An exemplary result in case of a single vehicle is shown in Fig. 6.5. The illustration is also referred to as a range-velocity map (sometimes also range-Doppler map).

Automatic target detection with a constant false alarm rate. The objective of an automated target detection is to locate peaks in the range-velocity map and to determine the distance as well as the velocity of the target according to the respective bins. Subsequently, the angle of the detected target is estimated using a third Fourier transform. Thereby, the peaks often distinguish only slightly from the noise floor, hence a primitive threshold-based method would lead to either too many false positives or false negatives. To keep the number of false positive detections rather constant, a separate threshold is calculated for each cell. Such algorithms are therefore also referred to as constant false alarm rate algorithms, abbreviated as CFAR.

For this purpose, a separate noise level is estimated for each cell of the range-velocity map based on the surrounding cells, similar to a sliding-window method. The noise level

[2]An alternative interpretation is based on the phase shift of the incoming signal due to the motion of the target. In this context, the beginning of the subsequent chirp travels a slightly different distance, resulting in a phase shift. Regardless of which interpretation is used, the actual Doppler shift is just estimated.

depends on the environment, thermal fluctuations and numerous other influences. To determine the threshold value Θ for the cell under test, a statistical model for the signal noise is used. The cell under test in column i, row j is represented by a stochastic random variable $X_{i,j}$. Similarly, the surrounding cells within a $n \times m$ sized window are represented by $X_{i+l,j+k}$ for $|l| \leqslant {}^{n-1}/_{2}$ and $|k| \leqslant {}^{m-1}/_{2}$.
If the signal power of the cell under test (due to a target object or a random disturbance) exceeds the threshold Θ, a target is detected and captured in the detection list (point cloud). In summary the general target detection procedure d is described by

$$d(X_{i,j}) = \begin{cases} \text{target}, & \text{if } X_{i,j} \geqslant \Theta \\ \text{no target}, & \text{if } X_{i,j} < \Theta \end{cases}. \tag{1.36}$$

In the case where no radar target is present, the threshold $\Theta = \alpha\mu$ can be decomposed into a product of average noise power μ and a scaling factor $\alpha > 1$. The probability for a false positive detection is now given by

$$p_{\text{FP}} = p\left(X_{i,j} \geqslant \alpha\mu\right). \tag{1.37}$$

If one assumes that the noise within the range-velocity map is complex normally distributed and $X_{i,j}$ is determined by the complex absolute value, then $X_{i,j}$ is exponentially distributed [9].
In order to calculate the probability defined in equation (1.37), different procedures for the calculation of the average noise power μ are distinguished. If, for example, a cell-averaging procedure, abbreviated CA-CFAR, is used, an average value is taken from the surrounding cells. Since targets usually stretch across several cells, one introduces an additional $n_g \times m_g$ window of guard cells, which are neglected when calculating the cell average. Consequently, the cell average is given by

$$\mu = \frac{1}{|L||K|} \sum_{l \in L} \sum_{k \in K} X_{i+l,j+k}. \tag{1.38}$$

The sets L and K are the surrounding cells without the guard cells, i.e.,

$$\begin{aligned} L &= \left\{ l \in \mathbb{N} \;\middle|\; \frac{n_g+1}{2} \leqslant |l| \leqslant \frac{n-1}{2} \right\} \\ K &= \left\{ k \in \mathbb{N} \;\middle|\; \frac{m_g+1}{2} \leqslant |k| \leqslant \frac{m-1}{2} \right\}. \end{aligned} \tag{1.39}$$

Provided the assumption of complex normally distributed noise is correct, equation (1.37) can be further simplified. Therefore, one obtains

$$\begin{aligned} p_{\text{FP}} &= p\left(X_{i,j} \geqslant \alpha\mu\right) \\ \ldots &= \frac{1}{\left(1 + \frac{\alpha}{N}\right)^N} \end{aligned} \tag{1.40}$$

for $N = |L||K|$. Since the false positive probability depends only on the constants α and N, one can conclude that the mean number of false detections is equal to p_{FP}, and

Fig. 1.5. The principal idea of the Open Simulation Interface, a language for standardized simulation frameworks. A concise and well-structured framework allows to exchange single simulation components and provides compatibility among different suppliers and guarantees various opportunities for application (see also [13]).

thus constant. This justifies the name constant false alarm rate. For a given tolerable probability of a false positive detection, the scaling factor is given by

$$\alpha = N\left(p_{\mathrm{FP}}^{-1/N} - 1\right). \tag{1.41}$$

The presented example serves only to illustrate a possible CFAR algorithm. In practice, the cell-averaging principle is not designed for multiple objects. Other targets inside the reference window distort the noise estimation and increase the threshold value as a consequence. Therefore, an algorithm that should overcome this problem is based on ordered statistics (OS-CFAR, see [10]), or a combination of ordered statistics and cell-averaging (OSCA-CFAR, see [11]).
Regardless which CFAR technique is used, each algorithm provides an adapted threshold. If the threshold is exceeded by the cell under test, the resulting detection is added to the detection list (point cloud) after a subsequent angle estimation. The point cloud interface is essentially the basis for the data-based radar sensor simulation presented in this thesis. However, for the sake of completeness, it should be mentioned that in practice the interface contains further data, e.g., an estimate of the radar cross section, or an estimate for the existence probability of an actual radar target. For further information on the signal processing see [12].

1.3 Modeling and standardized simulation frameworks

The objective of the sensor simulation is the generation of synthetic radar data based on a fully virtual environment. Ideally, radar-specific properties of all objects and resulting influences should be considered. Since the physical principles of electromagnetic wave propagation are well known, an exact simulation of the field distribution is reasonable. However, such a precise simulation raises numerous challenges for experts and their solution is currently not in sight. On the one hand, there are the tremendous amounts

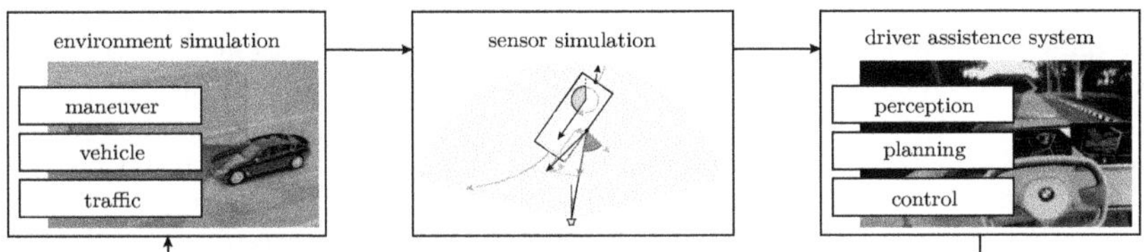

Fig. 1.6. Principle representation of the vehicle sensor simulation. The traffic and environment simulation provides an abstract, object-based description of the scene to the sensor model. The latter simulates the sensor-specific detection and supplies the data to the assistance system. The loop is closed by the executed driving maneuver.

of time and financial resources required to create a detailed realistic rendering of existing environments [14]. Regardless of whether these are urban or highway scenarios. On the other hand, such highly sophisticated replications of the virtual traffic world yield an unmanageably large number of fine structures and materials whose electromagnetic behavior cannot be fully simulated. In particular, such a sensor model would hardly be real-time capable due to the computational complexity.

A further dilemma, which is particularly encountered by car manufacturers, arises due to proprietary sensor software components. Once the manufacturer decides to equip the car with a certain sensor, usually only a few parameters are known. More detailed information, for instance about the antenna pattern, the wave form or the detection algorithms used, is not entirely disclosed by the sensor manufacturers. Therefore, from a car manufacturer's point of view, sensor simulation in the context of functional safety validation must be considered from a different perspective.

Many automotive driving simulation manufacturers thus only supply an interface to an abstracted description of the simulated world [15, 16, 17, 18]. Most of the environmental sensors in the automotive sector until a few years ago only handled an object list-based interface, so for historical reasons the description languages are inspired by that interface. Objects like vehicles, pedestrians and traffic signs are mostly described by two or three-dimensional surrounding cuboids (called bounding boxes) and supplementary parameters. Those parameters characterize for instance the dynamics, material specifications or other properties of the object. Likewise, the description language introduced by Hanke et al. and referred to as Open Simulation Interface (OSI) relies on this principle [19]. Furthermore, OSI is a reference implementation of the emerging standard ISO 23150 for sensor interfaces and uses in its current implementation the message format of the protocol buffers library developed and maintained by Google [20]. The idea is to connect any driving function with any simulation environment as easily as possible. This standardization reduces development costs, simplifies the integration of sensor models and strengthens the utilization of virtual tests. Essentially, the format can be reduced to two individual top level messages, defining the ground truth and the sensor data interface. A sensor model then uses the contents of the ground truth, executes a measurement model

and subsequently populates the sensor data through the standardized interface. Finally, the data can be retrieved by the driving function, the driver assistance system or simply the sensor fusion (after a possible conversion of the data) and interact with driving simulation (see Fig. 1.6).
Despite all advantages, standardization on the basis of such an abstraction level also poses certain challenges. While object-list-based sensor models are typically implemented using error models for the measurement deviation, new approaches must be developed to generate synthetic low-level data such as radar point clouds. Measurement models of radar sensors require not only physical principles, but above all numerous heuristics or data-driven approaches. Furthermore, model validation faces some unprecedented difficulties.

Chapter 2

State of research in automotive radar modeling

The first detection models of automotive radar sensors were introduced to enhance extended-object-tracking-algorithms [21]. However, due to the accelerated progress in the field of automated driving and driver assistance, this area has evolved into a field of research in its own right. Even more so, as safety validation and model deployment on HiL-test rigs places very individual demands on models. Thus, the contents of relevant publications in this field are summarized below.

2.1 Differentiation of various modeling levels

The environment simulation and sensor simulation framework have a decisive influence on the sensor modeling capabilities. The less details the environment simulation contains, the less effects can be covered by the virtual sensor. In order to compare models with different levels of abstraction, different modeling levels are usually distinguished. So far, however, the research community has not agreed on common terms.

In their work on the challenges of radar modeling, Holder et al. distinguish between ground truth, phenomenological, data-based and ray-tracing based sensor models as well as scattering center approaches [22]. Ground truth models are referred to as object list models. Phenomenological models may consist of statistical laws, simplified physical equations or maps to simulate typical sensor characteristics. While ground truth models mainly contain object lists, phenomenological models are still considered as their extensions. According to Holder et al., ray-tracing-based models use asymptotic approximations, i.e., geometric or physical optics, and provide raw data signals. Black-Box-Model is used synonymously for data-based models and provides synthetic, approximated sensor data based on real sensor measurements. Both deep generative networks and kernel density estimators are mentioned as examples [23, 24]. Here it should be objected, though, that such models by definition could also be classified as phenomenological, thus a sharp differentiation does not always seem possible. The category of scattering center approaches use a simplified object representation based on scattering centers in order to ensure fast computations. Again, it is important to emphasize that these models

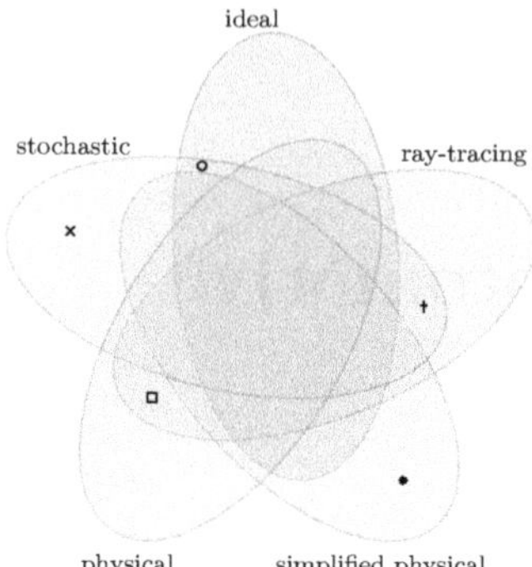

- ○ parametric sensor models presented by Hanke and various others [28, 29]
- × deep generative sensor models introduced by Wheeler et al. [23]
- □ ray-optical modeling approach of Hirsenkorn et al. [25]
- * abstracted radar measurement model of Berthold et al. [27]
- † two-dimensional ray-casting model introduced by Knill et al. [26]

Fig. 2.1. There are various categories for sensor models, but no common terminology. By means of a Venn-diagram and numerous examples, a possible model categorization shall be illustrated. In many cases it is not possible to assign models to only one category, so several overlaps are explicitly possible.

can be regarded as phenomenological, since they can also be construed as mathematical mapping.

In contrast, Hanke usually distinguishes between statistical (error) models and physical (measurement) models [30, 31, 3]. Although Hanke focuses mainly on a raw data format, the distinction between error and measurement model is in general quite convenient. Nowadays, though, the focus is moving towards unfiltered sensor data, which means that a steadily rising number of models use physical heuristics to emulate the measurement process. As a consequence, such a classification should refer to certain sub-components or -modules of a sensor model. Bernsteiner et al. did so in one of their earlier works on object list based models [32]. They consider sensor models as a composition of a geometrical (ideal sensor data), a phenomenological (real sensor properties) and a signal processing component (tracking).

A similar terminology like the one of Hanke is used by Hirsenkorn (but without explicitly distinguishing between error and measurement model), which in addition contains the category of deterministic sensor models [33]. By this, he means models based on mathematical mappings that consistently produce the same sensor data for identical object states and hence calls them deterministic. He briefly introduces neural networks as function approximators, but does not attach any practical relevance to them. Furthermore, the general categorization of neural networks as deterministic is no longer correct, since numerous neural networks can represent non-deterministic, generative models [34, 35].

In his work on the validation and calibration of sensor models, Schaermann mentions phenomenological and physical models [14]. He also calls the former models data-based, or statistical models. The physical models, which are also referred to as raw data models, include not only ray-optical approaches but also all methods that provide unfiltered sensor data. These include radar point clouds as well. Since Schaermann's naming is mainly

inspired by the sensor interface, the names might be misleading in some cases.
There are other publications, in particular the summary by Granström [21], which distinguish sensor models mostly on a point cloud format. They often differentiate between individual methods of statistical modeling and deal with physical approaches in the margins. In addition to ray-tracing approaches for lidar modelling, heuristic physical approaches are also considered as physical models.
Manufacturers of sensor simulation frameworks usually offer a variety of modeling options. This means, that complete object list based frameworks are provided, which can be categorized to the statistical error models according to Hanke, or the ground truth models according to Holder et al. [36]. Moreover, software suppliers of driving simulations sometimes offer ray-tracing interfaces, allowing experts to integrate their own ray-tracing models [37]. Also overall solutions of diverse categories are offered. Even though the naming of modeling levels may vary, these are mostly ground truth models (idealized models), statistical error models (HiFi - high fidelity sensor models) or raw data models (RSI - raw sensor interface models) [38]. Experts from car manufacturers have also contributed to the categorization of sensor models based on previous concepts [39].
As a consequence, neither the research community nor the industry have agreed on a uniform terminology. Since this work mainly deals with the simulation of automotive radar point clouds in standardized simulation frameworks, and therefore only a part of automotive sensor simulation, the following terms are used as lowest common denominator:

- **Idealized sensor models:** The most simplest models are deterministic and provide sensor data at an early stage of function development and also enable basic functional and safety tests. As an illustration, one can consider a model that assigns on or more radar detections to a reflective ground truth object. This may also include overlappings with simple ray-tracing based models.
- **Statistical sensor models:** Data-based and statistical models facilitate probabilistic modeling and/or use information extracted from real sensor data. This allows the modeling of measurement deviations with white noise, as well as the modeling of Poisson distributed false-positive targets, for instance. Typically, these models are all non-deterministic, whereas the effort required for implementation may vary significantly.
- **Ray-tracing based sensor models:** It seems reasonable not to identify the category of ray-tracing based models with a physical modeling approach. Particularly, since there are many different levels of abstraction for ray-tracing, ray-launching, and ray-casting approaches nowadays. Hence also models, which do not demand a high fidelity environment, belong to this category.
- **Simplified physical sensor models:** There are numerous models which reflect certain physical laws, but are also based on numerous heuristics. Usually, this kind of simulation does not solve the actual physical problem nor is it an approximation of it. In order to classify these models anyway, it is reasonable to call them simplified physical models.
- **Physical sensor models:** Now the category of physical models can be specified quite precisely and provides a simulation of the radar sensor technology as exact

as possible. This includes finite-difference-time-domain methods as well as other high-frequency approximations, if some sort of convergence is ensured. In practice, however, important material parameters are lacking most often. Furthermore, they often overlap with ray-tracing based modelling. The model introduced by Berthold et al., for instance, is not categorized as physical sensor model [27].

In order to make this categorization as generally valid as possible, intersections are not only feasible, they are even an explicit constituent as shown in Fig. 2.1. Due to rising detailed knowledge about the hardware and real sensor data, the modeling accuracy ideally increases throughout the development process [39]. This is also indicated by the above order.

2.2 Ray-tracing in environments of high-fidelity

Ray optical methods have been used for quite some time for high frequency simulation and especially for radar simulations. The idea to approximate the electromagnetic wave propagation with rays (in analogy to the fields of geometrical optics and the propagation of light) is used by Kouyoumjian et al. for calculating the scattering from dielectric bodies [40]. Moreover, numerous researchers have demonstrated similar techniques for the solution of partial differential equations [41, 42] and argued in favor of their applicability (see also [43, 44]). Despite this, numerous physical applications are mostly limited to simple scenarios with only a few objects. Complex ray optical simulations covering vast areas, such as those required in the automotive sector, are rarely available. Early on, Yun and Iskander stated that ray-tracing methods will be a game changer, though there are still challenges ahead [45]. Hence, first attempts in automotive radar sensor modeling using ray-tracing were also limited to small scenarios and less complex subproblems. The use of this technology was primarily focused on the simplification and improvement of complex simulations.
Schuler et al. used ray-tracing methods to extract scattering centers of vehicles. The objective of this was to use the scattering centers in order to reduce the high demands on the computing power of certain models. This allows the generation of simple car models for monostatic simulations, which in turn avoids the need of complex finite-difference-time-domain methods [46]. This was accomplished using a model based on the work of Maurer. However, this model does not take into account reflections on curved surfaces according to geometrical optics [47].
An analogous approach for the extraction of characteristic scattering centers was proposed by Buddendick et al. In this case, a high-frequency asymptotic field prediction tool is used to reduce the effort for ray-tracing models based on the shooting and bouncing ray principle [48].
With the integration of OptiX™ into commercial driving simulators ray-tracing became an increasingly important tool for the simulation of automotive sensors (see [49] for additional information). The first stable version of the user friendly ray-tracing interface was released in 2009 and has already been used in a wide variety of studies and simulations for the emulation of time-of-flight sensors [50, 51, 52]. After all, it was also used for the simulation of intermediate frequency signals of frequency modulated continuous wave

radar sensors. As an example, Hirsenkorn et al. introduced a shooting and bouncing ray based approach. Each ray represents thereby a cone, which expands in case of a reflection on a curved surface [25]. However, the curvature of each surface is uniformly defined in advance and is not based on the geometry of the object. Conscious that this is at best a rough approximation, the authors demonstrate that subjectively realistic results can be obtained in simple setups. Thus, this approach ensures that even in case of an extremely limited number of 30,000 rays, a sufficient amount of beams is received, and this while being real-time capable even without high performance GPUs. A decisive drawback of radar simulations using ray-tracing in high-fidelity environments might be the motion of objects. It is therefore possible that a ray-tracing scene is purely static and thus a simulation of the Doppler effect is not feasible a priori. This is probably one of the reasons why the approach suggested by Hirsenkorn et al. only simulates a single chirp. Thus, it only simulates the range measurement and hence cannot be used appropriately in driving simulations.

Meanwhile available driving simulation systems provide information about vehicle dynamics also through their ray-tracing interface. A model, which already uses the additional information regarding velocity, was presented by Holder et al. [53] and is similar to the proposal of Schick et al. [54]. In contrast to the previously mentioned approach of Hirsenkorn et al., they instantly simulate the radar cube, i.e., the range-velocity-angle map, instead of the intermediate frequency signal. Consequently, only a few parameters of the sensor (like measurement ranges, the number of bins, etc.) must be known, while computationally expensive operations such as the Fourier transform are avoided. Furthermore, experiments of the authors demonstrated that scattering centers are interpolated and a simulation of multipath propagation is possible. As is the case with any other model, realistic material parameters are lacking. And the fact that the measurement of those parameters is a demanding task has been demonstrated by the research of Kurz et al. [55]. Besides that, the question of convergence has not yet been fully clarified.

In their work on the simulation of range-velocity maps, Wald and Weinmann choose extremely simplified scenarios, but also a substantially different approach to provide a precise physical simulation [56]. To model the Doppler effect without relying on additional velocity information, a sequence of static scenes is used. For each measurement, a total of 128 consecutive sequences is generated during which the vehicle moves less than one tenth of the wavelength. Within the field of view there are more than 100 million rays emitted each time. However, these simulations are incredibly computationally expensive, meaning without GPU acceleration, the calculations take roughly an entire day. Therefore, the authors discuss an approach for the analytical evaluation of propagation paths, in which parts of the environment are simplified or neglected in order to shorten the execution time to less than a second. Unfortunately, a more detailed description of the actual technique is not included. Likewise, the accuracy of the fast method has not been clarified thoroughly.

2.3 Models executable in standardized environments

Car manufacturers are facing various challenges when it comes to the deployment of ray-tracing-based models in practice. Particularly relevant is the tremendous effort required for integration, accompanied by restricted interoperability due to the lack of standards

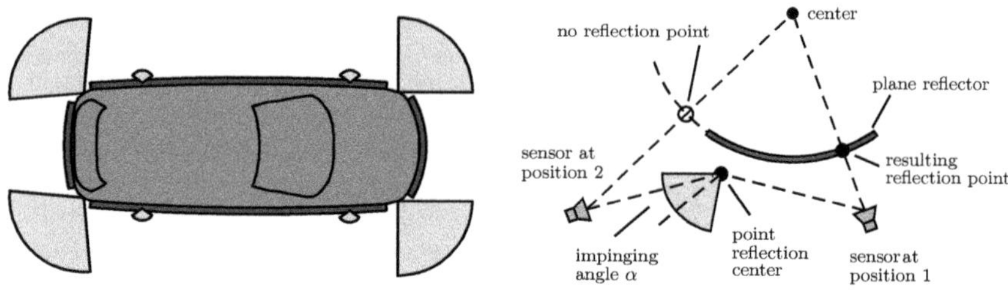

Fig. 2.2. Representation of a radar detection model proposed by Bühren et al. The model incorporates point-reflection centers and plane reflectors. While the point-reflection centers are ideal reflectors, the plane reflectors yield radar detections only if the electromagnetic wave impinges perpendicularly on the reflector (see [61]).

across vendors. Therefore, most interesting are those models which are in principle fully compatible with the Open Simulation Interface standard (see section 1.3).
There are already approaches addressing specific situations and phenomena. In regard to safety validation of driver assistance systems, for instance, weather influences are of increasing interest [57]. Also, disturbance effects due to exhaust gas clouds have been demonstrated already and are integrated into simulation [58]. The combination of multiple radar systems and the resulting interference effects must be increasingly taken into account [59, 60]. Despite all the research efforts in the area of simulation of isolated sensor effects, the focus in the following is placed upon the simulation of sensor systems under optimal conditions.
Accordingly, current technical knowledge in the field of simplified physical models (without the necessity of high-fidelity environments) will be discussed in the following section. Thereafter, current statistical and data-based models are discussed in depth.

2.3.1 Simplified physical models

Simplified physical models usually try to reproduce physical laws merely and therefore apply various heuristics. In order to be executable within standardized frameworks, the ground truth, i.e., a simplified representation of the scene, must be the baseline. Without having known about such standard, Bühren and Yang developed a model of a 24 GHz automotive radar sensor [62, 63] that could be deployed in such a framework. Similar to the later findings of Schuler et al. and Buddendick et al. [46, 48], they introduced point-reflection centers. These eight points, i.e., the corners and wheel cases, along with plane reflectors on each side are the cornerstone of their model. Point-reflection centers are considered as ideal detection if they are directly observable and the angle of incidence is within a predefined range (see Fig. 2.2). Ideal detections on plane reflectors, which represent the two-dimensional contour of the vehicle, arise precisely if a radar beam impinges perpendicularly upon them. Now the decisive point is that these detections form an ideal target list and not a radar point cloud as presented in section 1.2. Initially, each ideal detection is subjected with a measurement error for position, velocity, and angle. Afterwards, these noisy detections are used for the calculation of the signal amplitude. In case the amplitude within a resolution cell exceeds the 6dB threshold, a detection is generated. The model is tremendously useful practically, but often lacks detailed information about the resolution cells of the radar, or individual radar cross section values. Additionally, the modeling of the amplitude is contrary to the $1/R^4$-law and related research studies [33]. One reason for the linear decay of signal power observed by Bühren and Yang might be automatic gain control methods in the sensor technology used [64, 65]. Hence, their method cannot be adopted directly for other radar systems.

In a subsequent publication, Bühren and Yang presented extensions to their model [66]. In order to obtain a more realistic result for the amplitude simulation, a multipath propagation due to road reflections has been taken into account. Thus, radar energy, which propagates along indirect paths between sensor and target, e.g., sensor - road surface - object - road surface - sensor, and interferes with the waves of other paths, is now also included. Moreover, multiple reflections at short distance, which may cause ghost targets, are incorporated. The authors thereby refer to ping-pong reflections, i.e., sensor - object - ego vehicle - object - sensor. However, in their paper, they do not confirm these effects with real measurements. Moreover, the model was adjusted and used in various research studies [67, 68].

Another proposal completely deployable in an Open Simulation Interface framework, which has been developed before, was made by Knill et al. [26]. However, it is important to note that this model is rather a heuristic than a physical model. Starting from a simplistic ray casting approach, a model for approximately rectangular shapes in arbitrary traffic scenarios is introduced. At first, the intersection points of rays and objects are calculated in polar coordinates. Theoretically, up to two intersection points can exist along each ray and their measurement errors are modeled as follows: Angular errors are supposed to be uniformly distributed within a previously defined interval. The error involved in velocity measurements is assumed to be normally distributed. And lastly, the distance to the intersections is combined with a white noise model and thus yields a Gaussian mixture model for the range measurement (see Fig. 2.3). The model was used and tested in both single-object applications with a Rao-Blackwellized particle filter and in multi-object scenarios together with a labeled multi-Bernoulli filter [69].

Another, rather physically driven model was presented by Berthold et al. [27]. The radar's two-dimensional field of view is separated into range-angle resolution cells. Then, the energy flow in these individual cells is analyzed. The radar's two-dimensional field of view is separated into range-angle resolution cells. Then, the radial energy flow in these individual cells is analyzed. The entire energy is thereby sum of transmitted, reflected and absorbed energy. In this way, propagation along the underbody of the vehicle, but also through materials, can be taken into account. After the power analysis has been performed and an additional angular measurement error has been taken into account, the amplitude is modeled and a CFAR detection algorithm is applied. However, the algorithm is performed for each angle bin in radial direction and is contrary to the conventional range-velocity approach.

The model of Berthold et al. is based on previous research regarding the detection proba-

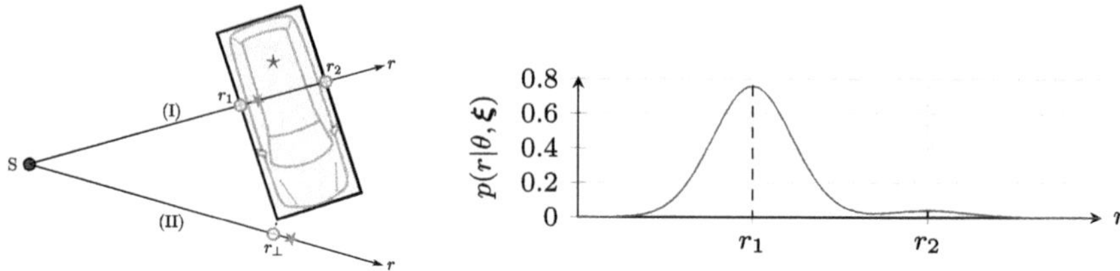

Fig. 2.3. The principal idea of the model introduced by Knill et al. Individual rays are emitted from the sensor. These can intersect with the vehicle's bounding box up to two times. The distribution for the position measurement (on the right) is obtained by a sum of weighted Gaussians. Even rays that slightly miss the target can cause a detection (see [26]).

bilities related to the contour of a vehicle [70]. Similar to the results of other researchers, the wheel cases were identified as a dominant feature detectable almost independently of the incident angle. With regard to the synthetically generated sensor data, however, this effect seems to be slightly overestimated by the model.

In a subsequent paper, they added further effects to the model [71]. A crucial aspect in this context is the integration of the radial velocity. In particular, the consideration of micro-Doppler effects, caused by the rotation of the vehicle wheels, is a significant progress in sensor modeling and will possibly be a decisive demand on sensor models due to future tracking algorithms (see [72, 73] for tracking-algorithms based on Doppler information). The velocity distribution is thereby modeled using a Gaussian mixture model and merged into the previously presented simplified physical approach. The parameterization is done with the help of real measurement data.

2.3.2 Statistical sensor models

Considered as separate models, the objective of statistical radar detection modeling is the reconstruction of spatial detection distributions for a given ground truth state (see section 5.1 for the mathematical problem formulation). In case of a constant number of detections per object, one also uses the term set of points on a rigid body. The general case is also referred to as spatial distribution models [21, 74]. For historical reasons, however, statistical models that operate on an object list level should be mentioned first. Even if they seem to be only marginally relevant for the problem statement, their approaches are essential.

Early approaches to automotive sensor simulation within driving simulators were developed by Rasshofer et al. [28]. By assuming a certain distribution for measurement deviations, purely idealized sensor models are made noisy. Furthermore, typical errors were identified, e.g., false negatives, false-positives, and latency, which were also modeled probabilistically with simple distribution functions.

Hanke et al. adopted this concept, formulated it in more mathematically precise terms and presented a modular architecture for such models [29]. Among statistical sensor models, these models are frequently referred to as parametric sensor models. Their name is derived from parametric statistics, which assumes probability distributions, e.g., Gaussian, Poisson, or Gamma [75]. In parallel (and in collaboration), Hirsenkorn et al. developed non-parametric modeling approaches without requiring additional assumptions on underlying probability functions [76, 24].

Previous point cloud models designed on the basis of probability distributions have not been applied in driving simulators. Instead, most of them have been used in conjunction with the development and improvement of algorithms for tracking extended objects. Thus, they have been integrated into more advanced tracking algorithms in form of model assumptions. The formulation, which is also important in the context of the actual sensor simulation, of the problem originates from Gilholm and Salmond [74, 77]. For tracking extended objects, they developed a Bayesian filter based on a probability model for radar point clouds. The authors refer to this model as extended target likelihood model. This does not yet represent a stand-alone radar sensor model, but these papers roughly outline the basics of sensor modeling on a point cloud level. Moreover, approaches like Gaussian mixture models are proposed. Further research papers on this topic mention similar assumptions based on spatial detection distributions, e.g., in [78, 79, 80].

With recent and rapid progress in the field of deep learning, more powerful generative networks have been introduced. Wheeler et al. trained such networks and used them to generate synthetic radar raw data in the form of range-angle maps [23]. Therefore, so-called conditional variational autoencoders are used [81]. This neural network attempts to learn the sensor measurement principle, including all measurement errors and deviations, using real sensor data and associated reference data (as a ground truth scene description). Subsequently, sensor data can be generated from ground truth data, which are also subject to stochastic fluctuations. Decisive for the success in the training process is the metric that is used. The research by Wheeler et al. examined the usual variational autoencoder loss as well as an adversarial loss. In case of adversarial loss, one also refers to (conditional) generative adversarial networks [35, 82]. Besides, the use of Gaussian mixture models were also studied. In quantitative evaluation, the variational autoencoder trained with adversarial loss has proven to reproduce the radar range equation best. Evaluations performed on a root mean squared error basis indicated that variational autoencoder models are superior to Gaussian mixture models in this particular case. Despite their initial successes, the authors conclude that learning models suffer from major limitations in generalization to complex environments. And the amount of required data is not negligible either.

Conclusively, one may note that despite various statistical models, there is no known application on radar point cloud level so far.

2.4 Validation and verification of sensor models

Validation and verification techniques have been essential in various disciplines. This also applies to sensor modeling. The validation of sensor models is the basis of every sophisticated driving simulation and a prerequisite for the objective evaluation of subsequent driving functions. Therefore, the validation and verification of sensor models is also subject of numerous research projects [83, 13, 84]. Since the terms are used differently depending on the research community and their definitions occasionally diverge [85], this thesis adopts the definition of the Institute of Electrical and Electronic Engineers (IEEE), which are also used by the International Organization for Standardization (ISO). However, similar definitions can also be found in the work of Sargent [86]. In accordance with this:

- **Verification:** The verification of a computerized model substantiates that its implementation is correct. Moreover, the process ensures that it represents a conceptual model within specified limits of accuracy.
- **Validation:** The validation of a computerized model substantiates that a computerized model within its domain of applicability possesses a satisfactory range of accuracy consistent with the intended application of the model.

An additional definition, which is beneficial in many areas and especially for decision makers, is provided by Oberkampf and Trucano [87]. Therein, validation is considered as a process of determining the degree to which a model is an accurate representation of the real world from the perspective of the intended uses of the model. This is in slight contrast to the IEEE definition, which requires explicit substantiation of applicability. Since both definitions have their legitimacy in the context of sensor models, both definitions will be used synonymously in the following. Moreover, the validation process can be split into the following parts according to Sargent [86]:

- **Data validation:** Data validity is usually of secondary importance during validation. In most cases, sufficient data quality (to the best of ones knowledge and belief) is the prerequisite for successful operational validity and high model confidence.
- **Conceptual model validation:** Conceptual model validity refers to the analysis of underlying theories and assumptions. In the context of the models intended purpose, the model structure, its logic, and its mathematical and causal relationships are of particular interest.
- **Operational validation:** Operational validation ensures that the model behavior complies with the accuracy requirements in the context of the intended use.

Most of the validation, verification, and evaluation takes place in the latter part. Therefore, a wide variety of techniques can be used, e.g., parameter variability-sensitivity analysis and even turing tests. However, in the context of sensor model validation, methods based on a comparison of real and synthetic data are well established.
Since there are currently only few research publications on the topic of validation of automotive radar sensor models on point cloud level, the following also contains a brief overview of related work in neighboring research areas. In particular, related publications in the area of validation of object list based sensors and lidar point clouds are discussed. One of the first contributions to this topic is made by Roth et al. [88] and deals with the evaluation of sensor models on an object list level. During the development of Virtual Test Drive, a validation architecture was designed which has the capability to evaluate arbitrary models in arbitrary scenarios. The work is strongly application-oriented and covers topics related to semi-automatic scenario generation, specification of interfaces, and

semi-automatic comparison of real and synthetic data. However, no metrics or validation concepts are introduced.

In their work on the validation of vehicle environment sensor models Schaermann et al. propose a two-step decision process [89]. As already proposed by Roth et al., on the one hand real and synthetic sensor data are compared with each other (first step). Additionally, the results of subsequent fusion algorithms using occupancy grids are integrated into the validation process (second step). Suitable metrics are presented for the first time. The quantitative results of a cross correlation coefficient and the absolute overall error are presented. The substantiation of validity is explicitly not provided, nor is a degree of validation given. Indeed, the authors emphasize that there is still no certainty what metrics are best suited for the evaluation. Furthermore, they describe their tool chain in equal detail. In a subsequent publication, an improved lidar model by Hanke et al. is presented and tested in a static scenario based on the already known validation methodology [31]. In short form, these results are also presented in [37] by Schaermann and Hanke.

In their work on a generally accepted validation methodology, Rosenberger et al. emphasize the importance of suitable reference data. The relevance of the reproducibility of trajectories is also pointed out. Their proposed statistical validation methodology is based on preliminary work by Viehof [90] on the validation of vehicle dynamics simulations. Accordingly, a list of phenomena to be tested must first be identified. Subsequently, a sensitivity analysis is performed with respect to the specified phenomena in order to determine the relevant configurations and parameters. As already suggested by Viehof [90], in this way scenarios and maneuvers for which a falsification of the model is most likely can also be determined. Since an overall validation cannot be performed, this allows to focus on only a few, but relevant scenarios. Finally, the actual accuracies are determined. As previous authors did, Rosenberger et al. emphasize that the formulation of model requirements and the provision of metrics is due to a lack of experience a completely new discipline in the field of sensor data generation. The subsequent validation example involves a lidar sensor and follows Schaermann et al. regarding the choice of the metric (based on occupancy grids).

An initial approach for the evaluation of automotive radar sensors originates from Holder et al. [91]. Since no validity criteria for radar sensor models were known at that time and the degree of uncertainties as well as the number of detections is almost arbitrary, the authors propose a completely different approach to evaluation. A radar sensor model can be falsified if subsequent algorithms based on occupancy grid maps are not able to plan a similar path through free space as in case of real sensor data. Thus, initial discrepancies between synthetic and real data are even allowed. Experiments of the validation method were performed based on two test scenarios, which resulted in the falsification of model assumptions. Additionally, implications for further sensor modeling were derived from the results.

In consequence, the validation of sensor models is just at the beginning and therefore certainly not finalized yet. Schaermann defines in one of his further works important requirements for sensor model validation [14]. According to this and among other points, attention should be devoted to specification (similar to what Rosenberger et al. stated). Overall consistency is as important as traceability. Intuitiveness of the validation methodology is often neglected, but especially in interdisciplinary areas with numerous stakeholders, it is most relevant since this ensures a generall acceptance of the methodology. In addition to the work already listed, there is also an extensive review by Viehof and Winner on the topic of validation and verification, but without any direct reference to sensor modeling [92]. Although there has been a strong focus on operational validity, it should not be forgotten that the process of model validity without data validity generally does not guarantee reliable, satisfactory, and useful simulation results [93].

Chapter 3

Derivation of research questions, hypotheses and objectives

In the previous sections, numerous research efforts related to sensor simulation were outlined, and although some issues have already been resolved, several questions remain unanswered. Therefore, four research questions in the field of radar sensor simulations are formulated hereafter, which also form the backbone for the remaining thesis. Beginning with a thorough assessment of the modeling fidelity of an existing ray-tracing based physical model, the use of stochastic detection models based on generative neural networks will be motivated. The investigation and comparison of various stochastic models substantiates another approach, combining multiple modeling concepts. The validation issue remains essential for the assessment of future models, which is addressed in the last section.

3.1 Inaccuracies of current ray cone tracing based models

According to many authors, ray-tracing models will become increasingly valuable for radar sensor simulations [43]. However, the high demands especially on real-time feasible simulations cause significant limitations. Besides, in some modeling approaches, decisive radar phenomena are still missing, e.g., the influence of the Doppler effect in Hirsenkorn's simulation approach [25, 33]. As Wald and Weinmann properly pointed out in [56], the Doppler spectrum can be rebuilt with multiple consecutive frames, but this approach is hardly real-time capable. Therefore, Doppler simulation based on successive frames cannot be considered as an appropriate extension for the model of Hirsenkorn et al. Based upon the work of Holder et al. [53] and the fact that the OptiX™ interface of Virtual Test Drive also provides dynamic information, one can assume the method of ray-tracing based simulation of frequency modulated continuous wave radars can in principle be extended by a simple Doppler effect. However, it remains uncertain how to treat micro Doppler influences. As it is conceivable to avoid this using various heuristics, there is still the question of the reliability of the presented approach (especially concerning the inaccuracy of the beam expansion due to reflections). Since this is directly related to the number of radiated rays, there is also a need to clarify whether and to what extent a lower bound for the number of rays can be determined and how to further enhance the existing model in the future. In summary, the first research question can be stated as follows:

> *What is the fidelity of the ray cone tracing based model of a frequency modulated continuous wave radar sensor and what refinements are required prior to considering it for the generation of synthetic training data of a stochastic radar model?*

3.2 Stochastic radar models based on deep generative networks

For a long period of time, the focus of sensor simulation for test and safety validation of driver assistance functions was on an object list level. As development efforts in the field of automated driving, especially sensor fusion, increased, the demand for sensor technology with powerful imaging capabilities has emerged. In other words, a maximum number of reflections per object is needed [94]. As a consequence, adequate modeling approaches for simulation of radar point clouds were missing at first. The few known methods originated mostly from the field of algorithm development for extended object tracking and thus developed with another purpose in mind. Hence, the simulation of radar points clouds is restricted to simplified physical models in combination with various continuous false alarm rate alorithms. The approaches used are usually based on ray-tracing methods of varying complexity and without emulating a specific sensor as precisely as possible, though. Apart from that, there are already first statistical approaches using deep generative networks [23], which are supposed to achieve this effect through a learning process. For the moment, however, its use is limited to range-angle maps and is therefore of little relevance to automotive manufacturers and their sensors. Further statistical methods based on kernel density estimators, have been used for the simulation of measurement deviations on an object list level as well as for the generation of whole lidar point clouds [76, 24, 95]. Such modeling techniques are of significant value for automotive manufacturers and their radar sensor simulations. On the one hand, it is possible to use existing data from field testing for training and parameterization, whilst on the other hand, very little detailed knowledge about the sensor software and hardware is involved. Latter details are usually only known by the sensor supplier anyway.

Therefore, classical, statistical methods as well as methods based on deep generative neural networks shall be developed to enable a data-based simulation of radar detection lists. Thereby and above all, an unbiased benchmarking of all results is desired. Furthermore, it must be clarified how these methods perform regarding their generic nature. In other words, how purposive are these methods and what are their convergence properties? In summary, the second research question can be stated as follows:

> *Are statistical methods based on kernel density estimation and neural networks suitable for a data-based simulation of radar point clouds and how are their*

learning capacity and practicality to be assessed in general and in comparison to each other?

3.3 Hybrid multipurpose approaches for radar sensor models

A purely data-based, statistical simulation usually intends to learn the distribution functions of radar detections as accurately as possible. As a consequence, the results of the learning process might even be transferred to hitherto unseen scenarios. Even without being aware of the full potential of these methods, the approach raises numerous additional questions. For instance, it remains unclarified which modeling approaches to work with at earlier stages of development, i.e., up to the point when real sensor data are available. To what extent does a learning model retain its predictive power? What amount of real data is needed in order to properly train models, and does this not even exceed the actual safety validation effort?

For the latter question, Schaermann has already presented an interesting analysis [14]. Assuming at least one data point for each resolution cell of the radar, the recording time already exceeds by far half a year. And this already holds for ordinary radar systems. Although such a large number of data points is not compulsory, this already demonstrates the actual problem and raises serious concerns about the practicability of such concepts. A further decisive drawback of learning algorithms in the field of sensor modeling is their limited suitability for hardware-in-the-loop applications. Since the focus is often on functional testing under certain circumstances. As a relevant example, the blind spot assistant in conjunction with an approaching motorcycle is worth mentioning. The typical question about the functional performance in case of partial detection losses (false negatives) [59] cannot be answered with a learning model in general. Instead, without a specific parameterization, such statements can only be tested statistically, provided that such situations are present to some extent in the training data.

To avoid as many of these difficulties as possible, it is necessary to have models that combine several methods. Therefore, a modeling concept shall be developed which makes use of heuristical, respectively physical, as well as data-based approaches in order to be reasonably applicable in as many stages of development as possible. Only this way allows the model to be optimized using real data, which is subsequently obtained throughout the development process. In case of any radar sensor, the intention shall always be an exact simulation of spatial detection distributions on dynamic objects whenever possible. Previous work has focused mostly on simple underlying heuristics [26] or physical principles [27, 53], without emulating a sensor based on real data. In order to ensure selective function validation, the developed approach should also incorporate the possibility of parameterization for worst-case scenarios. The underlying heuristic, respectively simplified physical principles shall be the scientific backbone, ensuring the generation of "plausible" data even in case of an insufficient data basis. In regard to this, Cao et al. also refers to grey-box models, which, however, are based on physical wave propagation instead of general underlying heuristics [96, 97]. In summary, the third research question can be stated as follows:

Is there a modeling approach for radar point clouds in standardized simulation frameworks that is applicable in several stages of development, based on simplified physical principles, but also adjustable through the use of real sensor data and remains parametrizable and extendable?

3.4 Deficiencies of current validation criteria

The operational validation of automotive radar sensor models is currently still in its very early stages. Previous approaches, as listed in section 2.4, often refer to the validation of object list based models or lidar point clouds and cannot be directly transferred to radar point clouds. On the one hand, this is related to strongly fluctuating number of detections and, on the other hand, it is still unclear which metrics are suitable for the actual evaluation.

Further issues and deficiencies of the evaluation on an object list level is the association of sensor and reference data. It may happen that the sensor provides multiple targets for a single object. A subsequent comparison requires association algorithms in order to apply a metric. The situation is similar in the case of target failures. Even if a metric-based assessment is feasible, the question of the quality of the association algorithm used still remains. Hirsenkorn [33] and Schaerman [14] present similar algorithms for this purpose in their elaboration. However, an evaluation of the algorithms is missing.

In the case of model validation of radar sensor models on a point cloud level, the association issue is significantly simplified due to untracked information. However, the challenge remains to determine suitable metrics and quality measures in order to evaluate highly dispersed data and high uncertainty. As already mentioned by Holder et al., there are currently no research studies available regarding this topic [91]. The top priority in sensor model validation at a point cloud level is therefore the development of an appropriate quality measure (not necessarily a metric in the mathematical sense). Since previous validation attempts, especially the ones concerning lidar point clouds, only enumerate potential metrics and measures, the problem of a meaningful threshold definition requires special attention. After all, the mere possibility of applying metrics to point clouds does not yet provide a foundation for validation decisions. The often mentioned concept of intuitivity [14] has been neglected so far and shall be included in a performance measure for the spatial distribution of radar point clouds. In summary, the fourth research question can be stated as follows:

Is there a validation scheme for the evaluation of radar point cloud models that is in principle appropriate to take into account the statistical nature of automotive radar sensor technology, provides an intuitive and interpretable result beyond simple threshold definitions, and, in the case of an evaluation of real data, reliably classifies it as valid?

Chapter 4

Modeling challenges related to ray cone tracing

Chapter 2 has already provided a detailed overview of research activities in the field of radar sensor modeling based upon ray-tracing. Several authors assume the increasing relevance of this technology for radar sensor models in the future. On the one hand, it is possible to use these models directly in the driving simulator and hence to provide synthetic sensor data more directly to the subsequent driver assistance systems. On the other hand, there are also considerations to use high-fidelity models for the generation of training data [23], as it has been done before in the field of image processing [98, 99, 100]. However, the accuracy of those models must always be considered in this regard. The generated data of raw data models, though, are usually assessed only in a rather quantitative way.
Until date and in the context of general safety validation there is only one physical FMCW-model providing an intermediate frequency signal. The precision of the ray cone tracing based model, introduced by Hirsenkorn et al. [25], will be investigated in certain aspects in order to gain insight into possible limitations. The research results are subsequently used to derive fundamental requirements for ray cone tracing radar sensor models. Finally, an outlook on possible extensions of the model is given.

4.1 The caustic distance and the angular beam expansion

One of the distinguishing features of the FMCW model presented by Hirsenkorn et al. is the beam expansion caused by a reflection at curved surfaces (see Fig. 4.1). On the other hand, by real-time model requirements, which can most easily be achieved with a decreased number of rays. A reduced number of initially emitted rays inevitably leads to a diminished number of subsequently received rays. Unfortunately, this obvious correlation is even non-linear in many cases. To be received at all, the ray must pass the receiving antenna such that the antenna is located inside the beam cone. Taking a beam expansion due to reflection at curved surfaces into account, the number of receiving rays

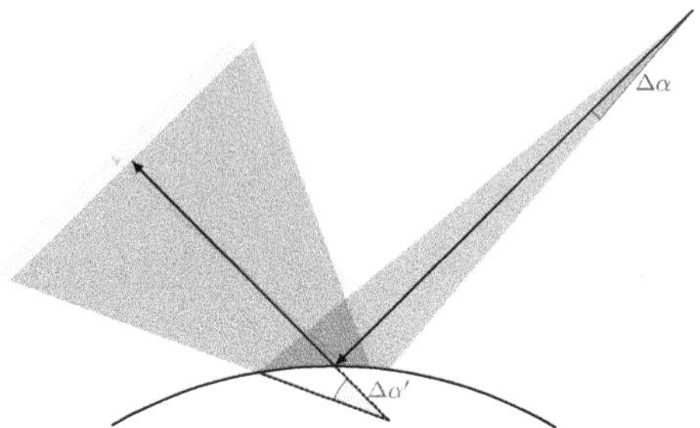

Fig. 4.1. Representation of the beam expansion due to the reflection on a spherical surface. Using one single ray in combination with an opening angle $\Delta\alpha$ the modeled reflection (blue) is only an approximation of the actual geometrical reflection (gray). The precision depends on the calculation of the caustic distance as well as the opening angle $\Delta\alpha'$ of the reflected beam.

can be increased significantly depending on the parameterization. With about 10,000 emitted beams, for example, only 0.05% are received by the antenna [25]. In case of 50,000 emitted rays already 0.11% [33]. One may assume that these numbers are further reduced if no additional beam expansion is performed and thus nearly no beams at all are received. However, the remaining question is if the approach is sufficiently accurate and geometrically justified.

If a ray cone is reflected at a curved surface, its opening angle changes. Provided that the cone is described by a single ray and a unique opening angle (as in the model used here), the point of impact is only an approximation of the point where the central ray of the reflected cone emerges from the surface (see Fig. 4.1). Additionally, the energy within the spherical cross-section of the cone is no longer uniformly distributed. As will be shown later, though, these approximation errors are considerably smaller than those of the approximation formulas used for the calculation of the opening angle and the caustic distance.

In case of perpendicular incidence, i.e., when the angle of incidence equals zero, a small-angle approximation formula is known from basic geometrical optics as mirror formula [101, p. 94 ff.]. In their work on scattering by dielectric bodies Kouyoumjian et al. have introduced an extended formula for the caustic distance [40][102, p. 435] which is equal to the mirror formula in the limiting case. For an arbitrary incident angle θ_0 and an incoming ray of length p the caustic distance q fulfils

$$\frac{1}{q} = \frac{1}{p} + \frac{2}{r\cos(\theta_0)}, \tag{4.1}$$

if, and only if, the illuminated surface is small, according to Kouyoumjian[1]. This condition is of course generally valid for ray based high-frequency approximations, but nevertheless

[1]Note that more general theories have already been developed and therefore further expressions for equation (4.1) already exist [103, 104].

not necessarily obvious. As will be shown later, the formulas are valid in a much more restricted range as is usual for small-angle approximations. For example, the error easily exceeds a 10% level for opening angles even less than one degree.
However, this fact was not known before and so Hirsenkorn et al. use the rough approximation of the caustic distance to estimate the opening angle $\Delta\alpha'$ for incident angles less than ten degree as follows

$$\Delta\alpha' = \frac{p}{q}\Delta\alpha = \left(1 + \frac{2p}{r\cos(\theta_0)}\right)\Delta\alpha. \tag{4.2}$$

To investigate the approximation errors in depth, a derivation of an exact formula for the spherical beam expansion in ray cone tracing based radar sensor models is given. The novel derivation is rather simplistic, since it requires only elementary knowledge of analytic geometry. Therefore, an arbitrary two-dimensional beam is characterized by two rays $v_{in}^{(0)}$ and $v_{in}^{(1)}$. Since the beam has a non-zero opening angle and it is reflected at a sphere one may assume

$$p = \left\|v_{in}^{(0)}\right\| < \left\|v_{in}^{(1)}\right\|. \tag{4.3}$$

Otherwise one can just switch their names. Without loss of generality the coordinate system can be chosen such that its origin is at the center of the reflecting sphere. Furthermore, the system is oriented such that the x-axis runs through the point of impact of $v_{in}^{(0)}$ (see Fig. 4.1 and 4.2). Thus, the impact point's coordinates are $(r, 0)$.
In the next step, one can compute the coordinates (x_+, y_+) of the second point of impact. It presupposes the parametrization of $v_{in}^{(1)}$ as a line. The inclination, i.e., the angle between the line and the x-axis, can be derived geometrically (see the right side of Fig. 4.2). In addition to the inclination, one needs another point on the line. The law of sines can be used to determine the distance between the first impact point and intersection point of the line with the x-axis. Hence the coordinates of the required second point of the line are given by

$$(r + l, 0) = \left(r + \frac{p\sin(\Delta\alpha)}{\sin(\theta_0 + \Delta\alpha)}, 0\right).$$

Finally, the straight line equation through (x_+, y_+) along $v_{in}^{(1)}$ is determined by

$$y = \tan(\theta_0 + \Delta\alpha)(x - r - l). \tag{4.4}$$

To find x_+ and y_+ in terms of the length p, the opening angle $\Delta\alpha$, and the incidence angle θ_0 one has to compute the point of intersection of the line with the two-dimensional sphere of radius r centered at the origin. Hence it is necessary to solve the quadratic equation

$$r^2 - x^2 = t_\alpha^2 (x - r - l)^2, \tag{4.5}$$

where $t_\alpha = \tan(\theta_0 + \Delta\alpha)$ was substituted for the sake of simplicity. The well-known formula for solving quadratic equations can be used to obtain

$$x_\pm = \frac{(r + l)\, t_\alpha^2 \pm \sqrt{r^2 (1 + t_\alpha^2) - (r + l)^2 t_\alpha^2}}{1 + t_\alpha^2}. \tag{4.6}$$

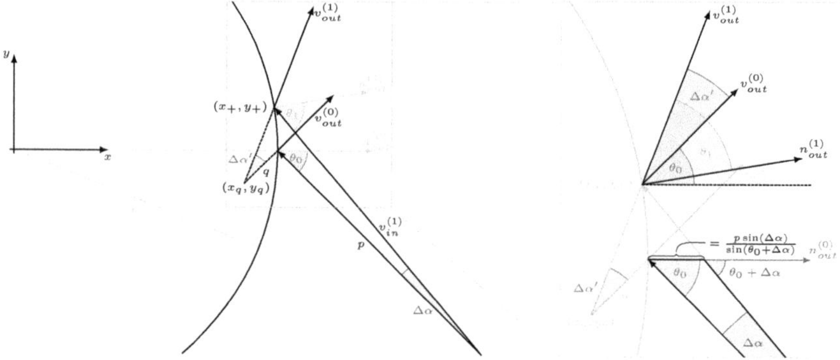

Fig. 4.2. Sketch for the derivation of an exact formula for the caustic distance and the opening angle of the reflected beam. Thereby, Fig. 4.1 is rotated by 90°. In order to clarify the interrelation $\Delta\alpha' = \theta_1 + \gamma - \theta_0$ the area around the impact point at $(r, 0)$ is shown in detail on the right side (with shifted vectors).

Due to the fact that $x_+ \geqslant x_-$, one can conclude that (x_+, y_+) is the first intersection point, provided that one starts from the origin of the beam. Thus, (x_+, y_+) is the only one relevant for further calculations. Accordingly, the y-coordinate of the impact point is given by

$$y_+ = \frac{(r+l)\,t_\alpha - \sqrt{r^2\left(1+t_\alpha^2\right) - (r+l)^2\, t_\alpha^2 t_\alpha}}{1+t_\alpha^2}. \tag{4.7}$$

The angle of incidence θ_1 between $v_{in}^{(1)}$ and the outer normal vector $n_{out}^{(1)}$, which corresponds to the position vector of (x_+, y_+) except for a normalization constant, is given by the dot product, i.e.,

$$\begin{aligned} -v_{in}^{(1)} \cdot n_{out}^{(1)} &= \left\|v_{in}^{(1)}\right\| \left\|n_{out}^{(1)}\right\| \cos\left(\theta_1\right) \\ \Leftrightarrow \qquad \theta_1 &= \cos^{-1}\left(\frac{x_+ - y_+ t_\alpha}{r\sqrt{1+t_\alpha^2}}\right). \end{aligned} \tag{4.8}$$

It is important to note that the negative sign of $v_{in}^{(1)}$ is required since $v_{in}^{(1)}$ is oriented towards the sphere and therefore the angle calculation would be incorrect otherwise. Now it is possible to deduce a closed analytic formula for the opening angle $\Delta\alpha'$ of the reflected beam. Referring to Fig. 4.2 from a purely geometrical point of view one observes that $\Delta\alpha' = \theta_1 + \gamma - \theta_0$ with $\gamma = \tan^{-1}\left({}^{y_+}/{}_{x_+}\right)$ and hence

$$\Delta\alpha' = \cos^{-1}\left(\frac{x_+ - y_+ t_\alpha}{r\sqrt{1+t_\alpha^2}}\right) + \tan^{-1}\left(\frac{y_+}{x_+}\right) - \theta_0. \tag{4.9}$$

To further derive an exact formula for the caustic distance q one needs to compute the intersection of the two exiting rays. In order to parametrize two straight lines through $(r, 0)$ in direction of $v_{out}^{(0)}$ and through (x_+, y_+) in direction of $v_{out}^{(1)}$ it is necessary to determine their angle of inclination. Obviously, as a consequence of the choice of the coordinate system the inclination of $v_{out}^{(0)}$ is equal to θ_0. A simple geometric consideration yields an inclination angle of $\theta_1 + \gamma$ for $v_{out}^{(1)}$. Again, in order to find the coordinates of the theoretical origin (x_q, y_q) of both rays one needs to compute the intersection of both straight lines, which is given as the solution of the linear equation

$$\tan(\theta_0)(x - r) = \tan(\theta_1 + \gamma)(x - x_+) + y_+. \tag{4.10}$$

Subsequently, by inserting x_q into the right-hand side of the previous equation, one also obtains y_q, i.e.,

$$\begin{aligned} x_q &= \frac{r\tan(\theta_0) - x_+\tan(\theta_1+\gamma) + y_+}{\tan(\theta_0) - \tan(\theta_1+\gamma)} \\ y_q &= \tan(\theta_0)\frac{(r - x_+)\tan(\theta_1+\gamma) + y_+}{\tan(\theta_0) - \tan(\theta_1+\gamma)}. \end{aligned} \tag{4.11}$$

Finally, since having (x_q, y_q) in terms of p, $\Delta\alpha$, and θ_0 it is possible to compute the caustic distance q as

$$\begin{aligned} q &= \mathrm{dist}((x_q, y_q), (r, 0)) \\ &= \frac{(r - x_+)\tan(\theta_1+\gamma) + y_+}{\tan(\theta_0) - \tan(\theta_1+\gamma)}\sqrt{1 + \tan^2(\theta_0)}. \end{aligned} \tag{4.12}$$

Although the formulas just derived can be expressed concisely, in terms of p, $\Delta\alpha$, and θ_0 the expressions are more than just elongated and complex. Consequently, equations (4.1) and (4.2) may still constitute more viable approaches in real-time applications than equations (4.9) and (4.12). But then, the scope of validity must be taken into account.

4.2 Estimating current errors in case of multiple reflections

In order to investigate the approximation error of current ray cone tracing-based radar sensor model implementations it is convenient to consider the relative error. With a view to equation (4.1) and (4.12) the relative error of the caustic distance q is given as

$$\varepsilon_q = \left|\frac{q - q_{\text{approx}}}{q}\right|. \tag{4.13}$$

A similar formula holds with respect to (4.2) and (4.9) for the relative error of the opening angle $\Delta\alpha'$ of the reflected beam, i.e.,

$$\varepsilon_{\Delta\alpha'} = \left|\frac{\Delta\alpha' - \Delta\alpha'_{\text{approx}}}{\Delta\alpha'}\right|. \tag{4.14}$$

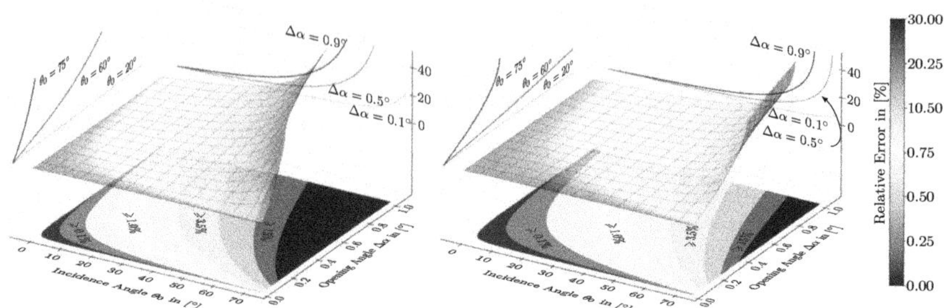

Fig. 4.3. Relative approximation errors according to the incidence angle θ_0 and the initial opening angle $\Delta\alpha$. On the left side the distribution ε_q for the caustic distance and on the right side the distribution $\varepsilon_{\Delta\alpha'}$ for the opening angle. In either case $p = 30$ and $r = 7$. To guarantee an error of less than 10% in previous implementations, but at the same time allow an incidence angle of up to 40°, the opening angle must be smaller than 0.6°. This value will be considerably lower for different model parametrizations, i.e., for highly curved surfaces.

The formulas (4.1) and (4.2) are actually small-angle approximations. A typical example of such an approximation is

$$\sin\alpha \approx \alpha, \tag{4.15}$$

which has a wide range of applications in branches of physics and engineering [105, p. 26 ff.]. In many cases a relative error of up to 1% is tolerated, meaning that the above formula is applicable for a range from −14° to 14°. However, relative errors of up to 10% might be acceptable in less sensitive applications. But analyzing the relative error ε_q and $\varepsilon_{\Delta\alpha'}$ obtained from equation (4.14) and (4.13), respectively, one observes at least two remarkable facts. On the one hand, both relative errors literally explode (see Fig. 4.3) and on the other hand, both approximation errors cancel each other out to a certain extent.

In this regard, the latter refers to the following fact: The approximation error arising from the calculation of the opening angle of the reflected beam calculated by means of equation (4.2) can be analyzed in two different ways. On the one hand, it is possible to analyze the error on the basis of the exact caustic distance q. On the other hand, one can also examine the error by using an approximated caustic distance q_{approx} according to the equation (4.1). The latter case is of course the more interesting and therefore also shown in Fig. 4.3. However, as already indicated, in some cases this error is much smaller than in the first case (cf. Table 4.1).

Although errors can be controlled easily, however, since ray cone tracing based models are often designed to simulate multipath propagation, multiple reflections have to be taken into consideration as well. Unfortunately, errors might grow exponentially, as the following example demonstrates clearly.

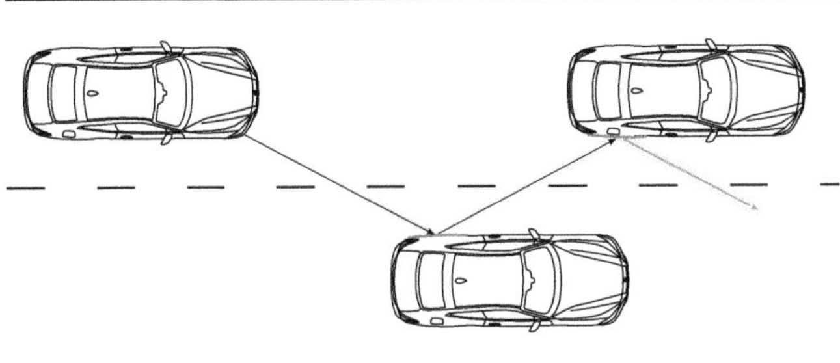

Fig. 4.4. Real world example of multiple reflections during a simple highway scenario. A beam hits the rear fender of a vehicle (the curvature is approximated with $r = 7\,\text{m}$ in red) in the adjacent lane at an incident angle of 60°. The beam is then reflected and hits the fender of the vehicle in front at an incident angle of 60°, again. In the end, the opening angle of the twice-reflected cone (represented by the blue ray) has an error of more than 10% when approximation formulas are used.

The ego vehicle is on the left lane of a highway. Two other vehicles are driving slightly offset to each other in front of it, one on the left and one on the right lane (see Fig. 4.4). Given a ray cone tracing based radar sensor model, one or more rays will move in such a way that they are reflected at the fender of both vehicles at an angle of incidence of 60°. Now, if the fender is spherically approximated with $r = 7\,\text{m}$, the calculation error of the caustic distance as well as the opening angle rises rapidly to more than 10% when using equations (4.1) and (4.2) (see Table 4.1). The error depends on the one hand, the distance covered by the rays (in this example about 5 m) and, on the other hand, the initial opening angle, i.e., the number of rays used.
However, it must also be mentioned that the model – as presented by Hirsenkorn et al. in [25] – does not yet consider a local approximation of the curvature. The approximation of the fender with $r = 7\,\text{m}$ is already close to optimal in this example. In general, though, the approximation is significantly worse, thus the actual error caused by those two equations can be neglected. Nevertheless, a locally spherical approximation is assumed in the following.
If, contrary to the example, the beam impinges on the rear right corner of the second vehicle and thus returns along the same path, the error is even higher due to significantly more curved surfaces. Under certain circumstances, this can even lead to opening angles of 90°. As a consequence, the beam is already received by the antenna as long as it is only slightly deflected in the direction of the ego-vehicle, i.e., if

$$-v_{\text{in}}^{(0)} \cdot v_{\text{out}}^{(0)} \geqslant 0. \tag{4.16}$$

Here $v_{\text{in}}^{(0)}$ is the initial emitted beam and $v_{\text{out}}^{(0)}$ represents the beam cone with an opening angle of 90°. Analogous to these two examples, there are countless scenarios in which

Table 4.1. Error propagation in case of repeated reflection analogous to the example from Fig. 4.4 with an additional third vehicle on the right lane. The lightning symbol indicates a large ray cone that illuminates more than the entire sphere. Angles of incidence $\theta = 60°$, curvatures $r = 7$, and initial distance $p = 5$.

	caustic distance			opening angle		
	q	q_{approx}	ε_q	$\Delta\alpha'$	$\Delta\alpha'_{\text{approx}}$	$\varepsilon_{\Delta\alpha'}$
			10,000 rays, opening angle 0.81°			
1st	1.21	1.30	7.47%	3.23	3.12	3.22%
2nd	0.79	1.37	73.40%	17.83	14.35	19.50%
3rd	↯	1.37	↯	↯	66.58	↯
			1,000,000 rays, opening angle 0.081°			
1st	1.29	1.30	0.67%	0.31	0.31	0.31%
2nd	1.33	1.37	3.00%	1.44	1.43	0.69%
3rd	1.15	1.37	19.10%	7.15	6.64	7.13%

similar effects occur. Therefore, if one does not consider the accuracy of the modeling algorithms in detail, one easily obtains results far from the intended objectives. In order to generate synthetic training data or to use ray cone tracing based models for safety validation purposes in the future, some geometrical aspects have to be considered.

4.3 Consequences and lower bounds for the number of rays

In order to be able to use ray cone tracing based sensor models meaningfully in a high-fidelity environment simulation, numerous further developments are required, as the current state of the art is still far from being sufficiently accurate. In addition to reflectivity parameters for various materials, which are not yet available, there are also geometric shortcomings. However, these shortcomings can be reduced by suitable model adaptation and a detailed error analysis.

The already mentioned issue of simple spherical surface approximation is crucial. Astigmatic tubes (see Fig. 4.5) in conjunction with two different radii of curvature yield a much more precise approximation. Thereby, the beam cone is represented by a direction vector, two opening angles, and a caustic. The radii of curvature are supposed to be given by the respective principal curvatures of the object surface at the point of impact. The method has been described several times [102, 103, 104] and has many applications. However, its implementation may be severely laborious in complex scenarios. Furthermore, it is necessary to investigate how curvatures can be determined efficiently beforehand to keep the computing time within an acceptable range. Real-time capability may no longer

Table 4.2. Error propagation in case of repeated reflection analogous to the example from Fig. 4.4. Now, however, the beam hits the second vehicle perpendicularly at the rear right corner and returns to the sender. Angles of incidence $\theta_{1,3} = 60°$, $\theta_2 = 0°$, curvatures $r_{1,3} = 7$, $r_2 = 1$, and initial distance $p = 5$.

	caustic distance			opening angle		
	q	q_{approx}	ε_q	$\Delta\alpha'$	$\Delta\alpha'_{\text{approx}}$	$\varepsilon_{\Delta\alpha'}$
			10,000 rays, opening angle 0.81°			
1st	1.21	1.30	7.47%	3.23	3.12	3.22%
2nd	0.42	0.46	9.52%	44.71	42.43	5.10%
3rd	↯	2.13	↯	↯	108.62	↯
			1,000,000 rays, opening angle 0.081°			
1st	1.29	1.30	0.67%	0.31	0.31	0.31%
2nd	0.46	0.46	0.07%	4.21	4.22	0.24%
3rd	0.62	1.33	114.50%	22.32	17.39	22.09%

be possible, but such a model might still be used to generate synthetic training data for statistical radar sensor models.
In order to achieve an optimum of precision and runtime in subsequent applications, the following considerations must be taken into account:

- **Curvature:** Which curvatures occur in the model and which of them are relevant for the scenario at all? In general, the stronger certain curvatures are, i.e., the smaller the radii of curvature are, the more beams have to be emitted by the sensor model. It may be possible to set a lower bound for the radii of curvature to prevent every detail from being fully considered.
- **Distance:** What is the maximum range of the sensor's field of view? The further away an object is, the larger is the area illuminated by a beam cone. The illuminated area must be significantly smaller than the local approximation based on the principal curvatures at this point.
- **Angle of incidence:** At what angles do the rays impinge? The angle of incidence has a direct influence on the size of the illuminated area. Therefore, it is advisable to set an upper limit for the angle of incidence, in order to prevent the beam from widening for angles exceeding this limit.
- **Number of reflections:** How many reflections should be considered by the model? The more often a beam is reflected, the more it may widen. If the opening angle exceeds one degree, in most cases the error is already too large to provide reliable results.

In the highway scenario from section 4.2, for instance, the errors from table 4.2 could easily be reduced to less than 1.8% with an appropriate number of rays. In fact, if one emits more than 100,000,000 rays semi-spherically, the resulting opening angle is smaller than 0.0081°). Unfortunately, the estimate is only valid for this particular case. In case of a short range radar, for example, a more general estimate can be obtained by assuming the following worst case. The beam expansion shall consider curvatures of $r \geqslant 0.25\,\text{m}$, incidence angles of $\theta \leqslant 60°$ and distances of $p \leqslant 30\,\text{m}$. If we now use 100,000,000 ray cones (emitted semi-spherically), even in case of a reflection at a distance of 30 m, the error is only about 3% at most and thus negligible. However, if up to three reflections are to be considered, whereby their radius of curvature is $r = 0.25\,\text{m}$ in the worst case, already 10^{15} beams are required to respect a 5% error threshold. These considerations show that computational resources required for precise simulation by ray cone tracing can reach astronomical dimensions quickly and can only be used in real-time with certain constraints. Although the field of view is usually much smaller and thus it is by no means necessary to emit the beams semi-spherically. Nevertheless, it impressively demonstrates the range in which the opening angles of the beam cones have to be. Even if exact formulas are used and consequently only small errors remain, large expansions can easily occur, which result in the area illuminated by the beams becoming too large.

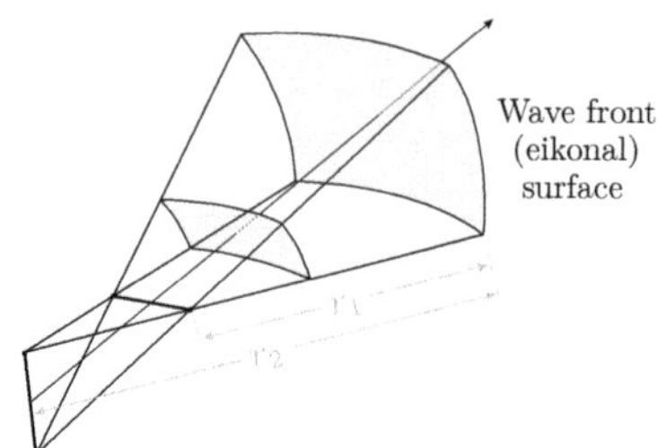

Fig. 4.5. Astigmatic tubes are more suitable than simple cones and may adapt better to the FoV.

Conclusion and recap of the first research question: At present, the existing ray cone tracing based model of a frequency modulated continuous wave radar sensor cannot be used for the generation of synthetic training data. On the one hand, the model has still obvious deficiencies, such as the missing emulation of the velocity and the angle characteristics. On the other hand, it currently lacks a large number of reflection parameters. In particular, the considerations of the previous chapter have shown the tremendous number of rays required for a sufficiently accurate simulation. This holds true even if astigmatic tubes are used to improve the approach in general. Consequently, in case of training statistical models, real data must be used instead of synthetic data.

Chapter 5

Approaches to statistical radar point cloud simulation

Early approaches to sensor simulations were already of a statistical nature. At that time, however, mainly focused on an object list level and without a systematic use of large data sets. It was particularly the previously discussed research of Hanke and Hirsenkorn (see chapter 2) which moved this type of modeling to the forefront and strengthened the use of real sensor data. The primary objective was the estimation of distributions of measurement errors and deviations in order to transfer the sensor behavior into simulation.
For a long time, statistical modeling concepts for the simulation of point clouds were only used as part of the algorithms developed for extended object trackers. However, the purpose here is completely different from the actual sensor simulation based on big data, thus these methods are only applicable to a limited extent for automotive manufacturers in the context of a safety validation of automated driving functions.
However, an important modeling approach was presented by Hirsenkorn et al. and is based on a kernel density estimation technique. In contrast to the classical methods of parametric statistics, this approach appears to be principally applicable to radar point clouds as well. The second method, which is in essence able to simulate radar point clouds, is based on neural networks and has already been used for the simulation of range-velocity maps.
After specifying the problem of sensor modeling mathematically more precisely, the following chapter generalizes these two approaches to detection lists. Finally, the learning capabilities of both methodologies are compared by means of a synthetic dataset and selected metrics (some of the results have already been published in [106]). Additionally, future application possibilities are discussed and assessed.

5.1 Statistical formulation of radar sensor modeling

The application of statistical methods in the field of radar sensor modeling first requires a mathematical formalism. In order to provide a thorough basis for the subsequent sections, a statistical formulation of the radar sensor model problem is given below, which is based on the well-known contributions of Hanke, Granström, and others [29, 21, 74].

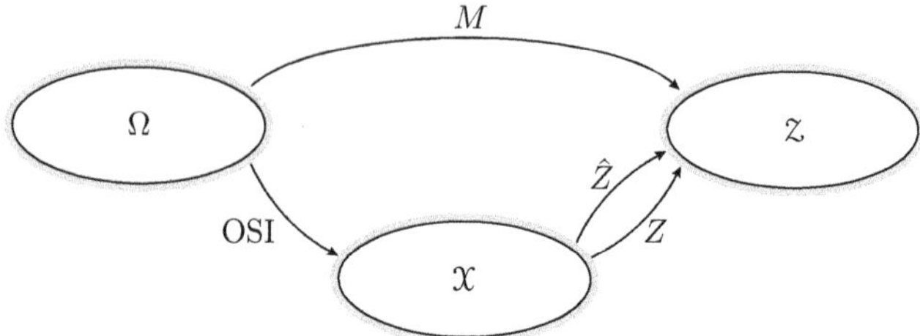

Fig. 5.1. Abstract illustration of the sensor measuring process in terms of mathematical mappings. The actual measurement process M operates on the entire reality Ω. Since the ground truth information $\mathcal{X}$ is logged and provided by a standardized interface, e.g., OSI, the sensor measurement principle is actually considered to be a mapping Z from $\mathcal{X}$ to $\mathcal{Z}$ only. Therefore, one needs to find a virtual sensor model $\hat{Z}$ which behaves similarly to Z.

The measuring process of a radar sensor is mathematically described as a mapping M. The abstract and almost elusive state of the environment is represented by an element of the set Ω. The result of the measurement process M is a radar point cloud $Z_{\text{real}} \in \mathcal{Z}$. In practice, one reduces the description of the environment to a standardized interface. This state, noted as an element of $\mathcal{X}$, can be considered as the result of a (standardization) mapping OSI itself. Hence, the measurement process reduces to a mapping Z. Consequently, the measurement process assigns a state X_{real} to a radar point cloud Z_{real} and hence it satisfies

$$\begin{aligned} Z\colon \quad \mathcal{X} &\to \mathcal{Z} \\ X_{\text{real}} &\mapsto Z_{\text{real}}. \end{aligned} \tag{5.1}$$

As depicted in Fig. 5.1, the simulated sensor is likewise realized by a mapping $\hat{Z}$ and maps a simulated state X_{sim} to a simulated radar point cloud Z_{sim}, i.e.,

$$\begin{aligned} \hat{Z}\colon \quad \mathcal{X} &\to \mathcal{Z} \\ X_{\text{sim}} &\mapsto Z_{\text{sim}}. \end{aligned} \tag{5.2}$$

The main objective of a sensor simulation is to approximate Z as precisely as possible by $\hat{Z}$. Of course, the information contained in $\mathcal{X}$ is decisive for the success of the simulation. An abstract description through a standardized interface inevitably leads to a loss of information.

Statistical modeling methods usually rely on real sensor data to simulate the measurement process. In most cases, this data is even labeled, i.e., additional reference sensors (a differential global positioning system combined with an inertial navigation system) are used to capture certain aspects of the environment. Since this work is mainly concerned with the spatial distribution of radar detections, the detection lists are limited to the target object and its immediate surroundings. Similarly, the description of the environmental state is limited to the relative motion of the target vehicles and consequently neglects additional effects. A labeled dataset consists of tuples of object states $X_{\text{sim}}^{(i)}$ and radar point clouds $Z_{\text{real}}^{(i)}$, i.e.,

$$\left\{\left(X_{\text{real}}^{(i)}, Z_{\text{real}}^{(i)}\right)\right\}_{i=1,2,\ldots,n}. \tag{5.3}$$

The ground truth state of the objects relevant to the sensor encompasses, for instance, the position, the velocity, the yaw angle, its spatial extents, etc. Since radar point clouds are independent of each other in time, the object state of the target vehicle appears to be completely random to the sensor. Hence, $X_{\text{sim}}^{(i)}$ itself can be thought of as a realization of a random variable X. If one assumes that the measurement result depends primarily on the state of the object, i.e., on $X_{\text{sim}}^{(i)}$ (as a rule, however, also additional effects, which are neither all known nor have been recorded, influence the result), then one is therefore interested in the conditional probability

$$p\left(Z = Z_{\text{real}}^{(i)} \mid X = X_{\text{sim}}^{(i)}\right). \tag{5.4}$$

Usually a shorter notation is preferred and one will simply write $p(Z \mid X)$ or $p_{Z|X}$. For modeling it is quite advantageous to restrict oneself to single detections instead of entire point clouds. If one considers the measurement process of a single detection $z_{\text{real}}^{(i,j)} \in Z_{\text{real}}^{(i)}$ as a random variable z as well, its likelihood is given by

$$p\left(z = z_{\text{real}}^{(i,j)} \mid X = X_{\text{sim}}^{(i)}\right) \tag{5.5}$$

and is of particular interest. In condensed notation one also writes $p(z \mid X)$ and $p_{z|X}$ once again. The conditional distribution $p(\cdot \mid X) = p_{z|X}(\cdot)$ is also called spatial distribution. If one suceeds in reproducing this distribution in its entirety for all conceivable states of the target vehicle, then a sensor model is realized by re-sampling. Obviously, it is crucial for this approach to determine the number of detections $n_{\text{real}}^{(i)} = \left|Z_{\text{real}}^{(i)}\right|$. Again conceptualized as a random variable n, the probability for it is given by

$$p\left(n = n_{\text{real}}^{(i)} \mid X = X_{\text{sim}}^{(i)}\right) \tag{5.6}$$

and also of particular interest. To avoid confusion, in case of doubt, the conditional distribution of the detection number is denoted by $p_{n|X}$ and the spatial distribution is denoted by $p_{z|X}$. Once the number of detections has been determined, the point cloud is generated by repeatedly sampling. In order to preserve a sufficient distance between different detections, it is recommended to use an acceptance-rejection method (see [107, p. 528 ff.] for a detailed description).

In practice, a sufficient data basis cannot ever be generated and thus the conditional distributions are only known in exceptional cases for a few states. Therefore, statistical sensor models have to reproduce the two distributions as exact as possible based on a limited amount of data. Afterwards, they should generalize even to states that have not been observed in the training data set before. This does not necessarily mean that the model is used for the simulation of entirely new states. However, a generalization is already necessary for even the slightest deviations in position or velocity. The following two chapters present two independent approaches to model the distribution functions $p_{z|X}$ and $p_{n|X}$ and also include the mathematical basics of their respective technique. The notation is consistently based on this section.

5.2 Kernel density estimation and radar point clouds

A fundamental problem of statistics is the determination of probability density functions and distribution functions. In parametric statistics, additional model assumptions are necessary. For example, one assumes that an n-element sample is the realization of n random variables, which are derived from a given family of probability distributions, and are uniquely determined except for certain parameters.
In contrast, no such assumptions are necessary in the case of non-parametric statistics. The model structure is therefore not a priori constrained and therefore best suited for the estimation of detection distributions on vehicles. A well-known method was presented by Parzen and Rosenblatt (independently) [108, 109] and already tested in the context of object-list based sensor modeling. The concept can be described at an object list level as follows (the example is essentially taken from the original work of Hirsenkorn et al. [24]): During a real measurement drive, the dataset

$$\left\{\left(X_{\text{real}}^{(i)}, Z_{\text{real}}^{(i)}\right)\right\}_{i=1,2,3,4} = \{(30, 30.5), (31, 32), (30.5, 30), (33, 33)\} \tag{5.7}$$

is recorded. Now, if one simulates a state $X_{\text{sim}} = 30.75$ that has never been observed before, the probability density function for the position measurement must be estimated according to the available dataset. Intuition suggests that known measurement deviations of nearby states are more relevant than the ones farther away. Additionally, one might assume the sensor measurements being normally distributed around the actual measured value (mathematically, this assumption is of no importance for the modeling success and hence not an additional model assumption as in case of a parametric approach). If one weighs the individual normal distributions with respect to their relevance, which is specified according to the similarity of $X_{\text{real}}^{(i)}$ and X_{sim}, and subsequently sums them up, one obtains an estimate of the probability density function $p\left(\cdot \mid X = X_{\text{sim}}\right)$ (see Fig. 5.2). Regardless of the measurement result, the measurements recorderd at $X_{\text{real}}^{(2)} = 31$ and $X_{\text{real}}^{(3)} = 31$ have a more significant impact on the estimation than the the ones logged at $X_{\text{real}}^{(1)}$ and $X_{\text{real}}^{(4)}$. That is just because of their smaller distance to the simulated state $X_{\text{sim}} = 30.75$. In the following, this principle is described mathematically and applied to two-dimensional spatial distributions of radar point clouds.

5.2.1 Principles of kernel density estimation

In its simplest case, the kernel density estimation procedure is used to approximate a density function f based on n observations $x_1, x_2, \ldots, x_n$. Depending on a strictly positive parameter h, also called bandwidth, the estimated density function is given by

$$\hat{f}_n(x) = \frac{1}{n} \sum_{i=0}^{n} K_h\left(x - x_i\right). \tag{5.8}$$

The function K_h is called kernel function and depends on h. As in the previous example, usually a Gaussian distribution is used for this purpose. However, arbitrary functions K_h

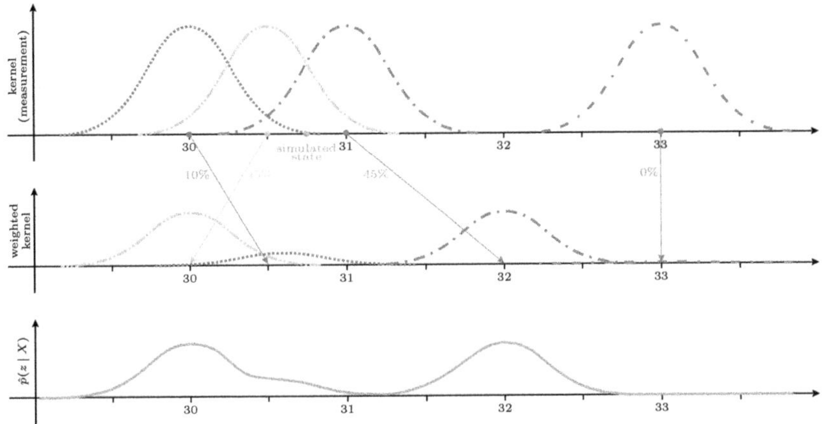

Fig. 5.2. Schematic illustration of the kernel density estimation technique based on the numerical example given in equation 5.7. A Gaussian distribution is drawn around each sensor reading (center of image). In dependence of the distance between the simulated state and the ground truth state at the time the data was recorded (top), the normal distributions are weighted. Subsequently, the estimated probability density distribution is given by summation (cf. [24]).

are conceivable as long as they fulfill the premises

$$\int_{-\infty}^{\infty} K_h(x)\mathrm{d}x = 1 \qquad \text{and} \qquad \int_{-\infty}^{\infty} xK_h(x)\mathrm{d}x = 0. \tag{5.9}$$

Common examples besides a normal distribution are the so-called Epanechnikov kernel and the tophat kernel. In general, however, the kernel density estimation procedure is relatively robust no matter which kernel is chosen [110]. Therefore, throughout this chapter a Gaussian kernel is used and equation (5.8) can be rewritten as

$$\hat{f}_n(x) = \frac{1}{n\sqrt{2\pi h^2}} \sum_{i=0}^{n} \exp\left(\frac{(x - x_i)^2}{2h^2}\right). \tag{5.10}$$

Clearly, the question arises whether the estimation given in equation (5.8) and (5.10) converges in a satisfying mathematical sense. According to Nadaraya's theorem [111], even weakest requirements ensure that the sequence of kernel density estimators $\hat{f}_n$ converge with probability 1 uniformly to the actual density function f, i.e.,

$$p\left(\lim_{n\to\infty} \sup_{x\in\mathbb{R}} \left|\hat{f}_n(x) - f(x)\right| = 0\right) = 1. \tag{5.11}$$

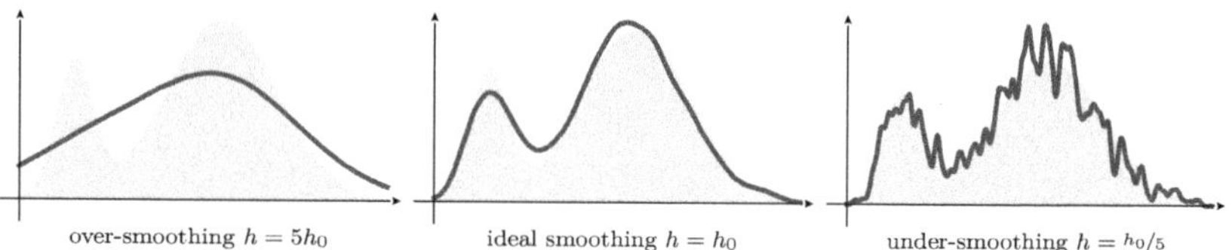

Fig. 5.3. Visual effects of different bandwidths on the quality of the kernel density estimation. A least square cross-validation method was used to determine the ideal bandwidth h_0.

Although, this result does not guarantee any good estimation in practice, it is crucial for non-parametric statistical sensor modeling. Indeed, the estimation principally gets better with a successively larger dataset. In particular, such a convergence statement does not exist for the deep generative neural networks discussed in the following chapter. Unfortunately, however, the theorem does not tell one how to choose the bandwidth. Consequently, this parameter is of superior importance for all practical applications and influences the quality of the estimation.

In case of a Gaussian distribution, the bandwidth is equivalent to the standard deviation, and thus the effects of over- and under-smoothing can be explained quite illustratively. If the bandwidth h is chosen too small, only individual peaks are summed up. The estimated density function fluctuates thereby considerably and it causes the so-called under-smoothing effect. If, on the other hand, the bandwidth is chosen too large, certain details based on local accumulations might be lost due to excessively wide normal distributions. This leads to so-called over-smoothing effect (see 5.3).

Therefore, the bandwidth h must be chosen in a way that the deviation between f and $\hat{f}$ is as small as possible. A suitable measure for the deviation is, for instance, the integrated square error, which corresponds to the squared distance of the metric induced by the L^2-norm, i.e.,

$$\begin{aligned} \mathrm{ISE}\left(\hat{f}_n, f\right) &= \left\|\hat{f}_n - f\right\|_2^2 \\ &= \int_{\mathbb{R}} \left(\hat{f}_n(x) - f(x)\right)^2 \mathrm{d}x \\ &= \int_{\mathbb{R}} \hat{f}_n(x)^2 \mathrm{d}x - 2\int_{\mathbb{R}} \hat{f}_n(x) f(x) \mathrm{d}x + \int_{\mathbb{R}} f(x)^2 \mathrm{d}x. \end{aligned} \tag{5.12}$$

The third term is independent of the bandwidth h. Consequently, the deviation between $\hat{f}$ and f is minimal if the first two terms from equation (5.12) are minimized. For a given bandwidth h, the first summand can be calculated easily. However, since the actual probability density f is unknown, the second term must be estimated. Thereby, this term can be interpreted as the expected value of $\hat{f}_n$ with respect to f and thus, can be approximated by simple averaging. In order to be able to apply a cross-validation procedure when minimizing, one disregards one sample value at a time. Noting the so-called leave-one-out kernel estimator with

$$\hat{f}_{n,-i}(x) = \frac{1}{n-1} \sum_{\substack{j=1 \\ j \neq i}}^{n} K_h \left(x - x_j\right), \tag{5.13}$$

for the second integral one obtains the estimation

$$\begin{aligned} \int_{\mathbb{R}} \hat{f}_n(x) f(x) \mathrm{d}x &= \mathbb{E}_f \left(\hat{f}_n\right) \\ &= \frac{1}{n} \sum_{i=1}^{n} \hat{f}_{n,-i}\left(x_i\right) \\ &= \frac{1}{n(n-1)} \sum_{i=1}^{n} \sum_{\substack{j=1 \\ j \neq i}}^{n} K_h \left(x_i - x_j\right). \end{aligned} \tag{5.14}$$

In order to minimize the unknown integrated square error ISE, one may minimize the function

$$\begin{aligned} \int_{\mathbb{R}} \hat{f}_n(x)^2 \mathrm{d}x - 2 \int_{\mathbb{R}} \hat{f}_n(x) f(x) \mathrm{d}x =& \frac{1}{n^2} \sum_{i=0}^{n} \sum_{j=0}^{n} \int_{\mathbb{R}} K_h \left(x - x_i\right) K_h \left(x - x_j\right) \mathrm{d}x \\ &- \frac{2}{n(n-1)} \sum_{i=1}^{n} \sum_{\substack{j=1 \\ j \neq i}}^{n} K_h \left(x_i - x_j\right) \end{aligned} \tag{5.15}$$

instead. This is usually done using a grid search algorithm together with the leave-one-out cross-validation procedure. Cross-validation was first proposed by Rudemo to determine the bandwidth [112]. The fact that this method is even asymptotically optimal was later proven by Stone [113].

5.2.2 Multivariate application to radar point clouds

In the multidimensional case, the bandwidth is represented by a positive definite matrix H. Together with a multivariate kernel K_H, the kernel density estimate in the multidimensional case is given by

$$\hat{f}_n(x) = \frac{1}{n} \sum_{i=0}^{n} K_H \left(x - x_i\right). \tag{5.16}$$

The bandwidth can be optimized analogously to the one-dimensional case with the help of a least square cross-validation. Using a multivariate normal distribution as kernel function, equation (5.16) can be rewritten as

$$\hat{f}_n(x) = \frac{1}{n\sqrt{(2\pi)^d \det H}} \sum_{i=0}^{n} \exp\left(-\frac{1}{2}\left(x - x_i\right)^\top H^{-1} \left(x - x_i\right)\right) \tag{5.17}$$

in the d dimensional case. However, now the spatial distribution of radar detections at a given simulation state X_{sim} shall be determined. In essence, the application here is analogous to the object list-based case where multiple measurement deviations are modeled simultaneously (cf. [24]). Therefore, one estimates the spatial distribution $p(z \mid X)$ from section 5.1 using Bayes' theorem, i.e.,

$$\begin{aligned}\hat{p}(z = z_{\text{sim}} \mid X = X_{\text{sim}}) &= \frac{\hat{p}(z = z_{\text{sim}}, X = X_{\text{sim}})}{\hat{p}(X = X_{\text{sim}})} \\ &= \frac{\hat{p}((z, X) = (z_{\text{sim}}, X_{\text{sim}}))}{c}.\end{aligned} \tag{5.18}$$

Here $c > 0$ is a normalization constant such that the integral of the density function equals one. In practice, multivariate kernel density estimation mostly uses a product kernel [114], which is equivalent to the use of a diagonal covariance matrix H. If one decomposes the bandwidth into two also positive-definite diagonal matrices H_1 and H_2, i.e.,

$$H = \begin{pmatrix} H_1 & 0 \\ 0 & H_2 \end{pmatrix}, \tag{5.19}$$

the probability density function from equation (5.18) can be simplified. In the particular case of a two-dimensional spatial distribution and a three-dimensional vehicle state (position and yaw angle, object dimensions are fixed), H_1 is a 2×2 matrix and H_2 is a 3×3 matrix. In essence the following applies to the kernel density estimation of the conditional density

$$\begin{aligned}\hat{p}(z = z_{\text{sim}}, X = X_{\text{sim}}) &= \hat{p}((z, X) = (z_{\text{sim}}, X_{\text{sim}})) \\ &= \frac{1}{N}\sum_{i=1}^{n}\sum_{j=1}^{m_i} K_H\left(\begin{pmatrix} z_{\text{sim}} \\ X_{\text{sim}} \end{pmatrix} - \begin{pmatrix} z_{\text{real}}^{(i,j)} \\ X_{\text{real}}^{(i)} \end{pmatrix}\right) \\ &= \frac{1}{N}\sum_{i=1}^{n}\sum_{j=1}^{m_i} K_{H_1}\left(z_{\text{sim}} - z_{\text{real}}^{(i,j)}\right) K_{H_2}\left(X_{\text{sim}} - X_{\text{real}}^{(i)}\right) \\ &= \frac{1}{N}\sum_{i=1}^{n}\sum_{j=1}^{m_i} w_i K_{H_1}\left(z_{\text{sim}} - z_{\text{real}}^{(i,j)}\right).\end{aligned} \tag{5.20}$$

The twofold summation occurs due to the fact that each ground truth state comprises several detections, i.e., there are m_i detections associated to the i-th state. The total number of detections is given by $N = \sum m_i$. Since z_{sim} is the only variable in the above equation, the kernel value of K_{H_2} can be calculated entirely, justifying the interpretation of the weighting from Fig. 5.2. An analogous model is implemented for number of detections, i.e., for the density function $p(n \mid X)$. The sampling of individual detections is done in two steps as suggested by Hirsenkorn [24]. First, one of the kernel functions is drawn based on the weights w_i. In the above notation, one must also consider that the multiplicity of the weights is related to the number of detections. Subsequently, one simply samples from the respective normal distribution. As mentioned before, in case of multiple sampling, a minimum distance between the individual detections may have to be taken into account. Therefore, the use of an acceptance-rejection method is recommended (see [107, p. 528 ff.]).

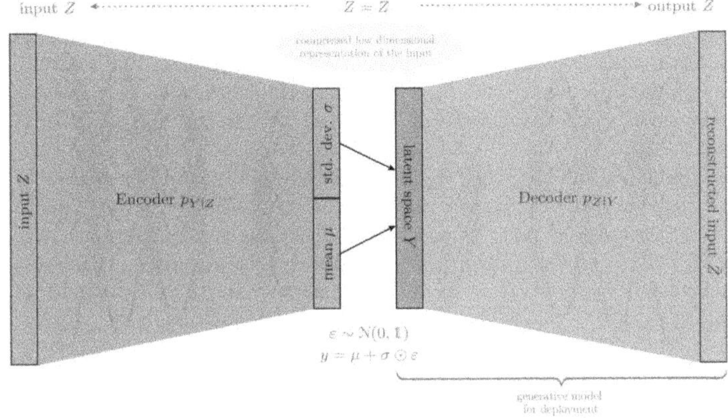

Fig. 5.4. The concept of a variational autoencoder (VAE) consisting of an encoder and a decoder. For mathematical details see also Fig. 5.5. Only the decoder is used as a generative model, while the encoder is required during training.

Another challenge is the choice of bandwidth in practice. The computing time of the optimization using a least square cross-validation already exceeds several days for larger datasets and is consequently not a scalable approach at all. However, this problem can be circumvented, since in general a globally optimal bandwidth is not necessary. However, since only ground truth states in the vicinity of X_{sim} exhibit a relevant weighting for the kernel density estimation, the computing time can be reduced significantly and locally an even better result can be achieved.

5.3 Deep generative networks as sensor models

Deep learning as a subfield of machine learning is of great importance nowadays. In particular, neural networks have been used very successfully for several years in computer vision, speech recognition, machine translation, medical image analysis, and many other areas.

The fundamental idea of the perceptron, the basic building block of any neural network, dates back to the late 1950s by Rosenblatt [115] and was subsequently extended by Ivakhnenko [116]. The first remarkable training successes can be traced back to LeCun, who trained neural networks using backpropagation and was thus able to automatically recognize handwritten digits [117].

The use of graphics processing units (GPUs) caused a revolution in deep learning. For the first time, convolutional neural networks dominated all other methods in a visual pattern recognition contest and became the dominant technique in computer vision [118].

At the same time, generative neural networks were developed by Goodfellow [35]. In

contrast to neural networks that recognize patterns and thereby learn decision functions, generative networks are used to generate new data, possibly never seen before. This approach has already been used to generate paintings, human faces, and video games. The results convinced even experts of these disciplines [119, 120, 121]. A similiar generative method based on neural networks was developed by Kingma and is called variational autoencoder [34].
As already mentioned in chapter 2, both approaches have already been used by Wheeler et al. for the generation of range-velocity maps. Based on this, the possible applications in the case of radar point clouds will be discussed in the following. Although the fundamentals of generative neural networkds are briefly presented, basic knowledge of neural networks is required. However, the basic principles can also be found in [122], for instance.

5.3.1 Variational Autoencoder

Autoencoders are actually used for efficient data encoding. For this, neural networks are often used and learn a low-dimensional representation of data in an unsupervised procedure. The compression from intial space into encoded space (also called latent space) usually entails a loss of information. In order to achieve the best possible reproduction of the input data by decompression, a maximum of information must be encoded and the reconstruction error must be decreased to a minimum. If a one-layer neural network with linear activation is used for both the encoder and the decoder, the principle can be thought of as a kind of principal component analysis.
To anticipate the idea of modeling radar point clouds with variational autoencoders, detection lists must be compressed and subsequently reconstructed during the training process. If one uses the decoder afterwards, a new and possibly unobserved, but still realistic radar point cloud can be generated by random sampling from the latent space. However, since this is not quite as simple as it sounds and also requires some intermediate steps, the basic principles will be briefly explained below.
In order to be able to use the encoded space, which entails the training data to some extent, for a sampling process, a certain regularity must be ensured. This guarantees that the encoded detection lists are distributed appropriately in the latent space. If this is the case, the space is also called well-organized and sufficiently regular for sampling. Therefore, variational autoencoder come into play. In simple terms, a variational autoencoder is an autoencoder with a latent space that is suitable for a generative process.
In contrast to an ordinary autoencoder, single data is not encoded as a point in the latent space, instead it is described in terms of a distribution with the parameters μ and σ. The re-parameterization ensures that the data in the latent space is standard normally distributed and thus the space is regularized.
Mathematically, this is accomplished as follows: During the encoding process[1], each data vector z is mapped to a random vector $y \in Y$. The distribution of the random vector can be chosen arbitrarily and is denoted by p_Y. In general, however, the distribution is a multivariate standard normal distribution. The encoder must therefore encode z by

[1]Usually, the latent space is denoted by Z. Since Z is already used for detection lists, the latent space is denoted by Y to prevent possible confusion.

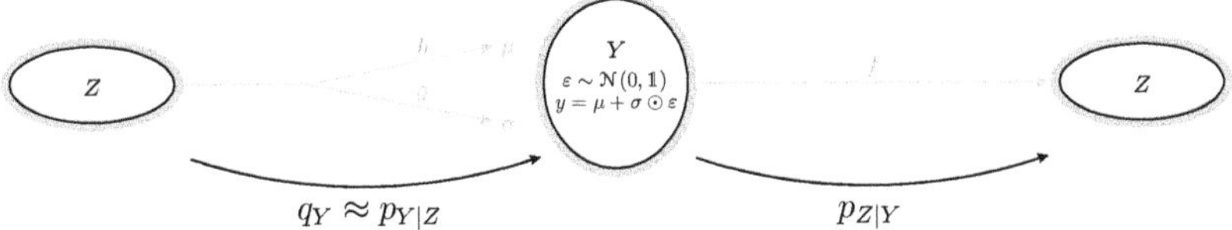

Fig. 5.5. Abstract mathematical representation of a variational autoencoder. The neural networks are represented by f, g, and h. The re-parameterization trick in Y ensures that the network can be trained by means of backpropagation.

the distribution $p_{Y|Z}$ (see Fig. 5.5). The decoder is subsequently defined canonically by $p_{Z|Y}$. If one assumes that p_Y and $p_{Z|Y}$ are normally distributed, the calculation of $p_{Y|Z}$ is a classical Bayesian inference problem. However, this condition is usually not fulfilled, hence approximation techniques are required. These methods are also called variational Bayesian methods and justify the name variational autoencoder (see [107, p. 462 ff.] for more information on variational Bayesian methods). Thereby the unknown distribution $p_{Y|Z}$ is approximated by a so-called variational distribution q_Y, i.e.,

$$q_Y \approx p_{Y|Z}. \tag{5.21}$$

The approximation by means of the distribution q_Y is also called latent variable model and is accomplished once again by means of a Gaussian distribution with the encoded parameters μ and σ. Therefore, the encoder corresponds to two functions g and h, which map the data vector z to the two parameters, while the decoder corresponds to a function f. An additional random vector $\varepsilon \sim \mathcal{N}(0, \mathbb{1})$ and the parameters μ and σ represent the actual random vector y in the latent space (see Fig. 5.5). The transformation is given by

$$y = \mu + \sigma \odot \varepsilon \tag{5.22}$$

and often called re-parametrization trick. An immediate training objective of the variational autoencoder is to obtain a satisfactory approximation of the unknown distribution $p_{Y|Z}$. The quality is determined by the Kullback-Leibler divergence D_{KL} (see [107, p. 55 ff.] for a detailed discussion of this quality measure) and must be minimized during training. For the approximation quality the relation

$$D_{\mathrm{KL}}\left(q_Y \,|\, p_{Y|Z}\right) = D_{\mathrm{KL}}\left(q_Y \,|\, p_Y\right) - \mathbb{E}_{q_Y}\left(\log\left(p_{Z|Y}\right)\right) + \log\left(p_Z\right), \tag{5.23}$$

introduced by Kingma can be derived easily [34]. If one re-arranges this equation once again, i.e.,

$$\log\left(p_Z\right) - D_{\mathrm{KL}}\left(q_Y \,|\, p_{Y|Z}\right) = \mathbb{E}_{q_Y}\left(\log\left(p_{Z|Y}\right)\right) - D_{\mathrm{KL}}\left(q_Y \,|\, p_Y\right), \tag{5.24}$$

one obtains the core equation of variational autoencoder. In fact, during training, the left-hand side of the equation (5.24) must be optimized. The term p_Z corresponds to the probability of generating a real detection list and is expected to be maximal. Conversely, the latent variable model $p_{Y|Z}$ should be approximated as good as possible, hence the second term on the left hand side must be minimized. Therefore, the variational autoencoder is optimal if the left-hand side equals zero. The expected value on the right hand side is also called maximum likelihood estimation and represents the reconstruction loss $\mathcal{L}$ which is usually a common loss function, e.g. the mean squared error. Consequently, the training objective is expressed by the optimization problem

$$\underset{f,g,h}{\arg\min}\, D_{\mathrm{KL}}\left(q_Y \,|\, p_Y\right) + \mathcal{L}\left(z, \tilde{z}\right). \tag{5.25}$$

The desired functions f, g, and h are approximated by neural networks. As usual, the whole variational autoencoder is trained by backpropagation, e.g., stochastic gradient descent. For a subsequent deployment as sensor model one only requires the decoder part, i.e., the function f. Given a random standard normally distributed vector from the latent space a new detection lists can be generated.
Although this principle is suitable for generating synthetic radar point clouds, the method itself does not yet guarantee detection lists that match the current simulation state. Consequently, the concept has to be extended. In this case one speaks of a so-called conditional variational autoencoder [81]. Both the encoder and the decoder receive the simulation state as additional input (cf. Fig. 5.6).

5.3.2 Generative Adversarial Networks

As already mentioned, the principles of generative adversarial networks were developed by Goodfellow et al. and originate from the field of computer vision [35]. In contrast to variational autoencoder, the mathematical description is considerably simpler. A generative neural network consists of two independent neural networks, namely the discriminator D and the generator G. The discriminator is a binary classifier which distinguishes between real and synthetic data. Thereby, it assigns a value in the interval $[0, 1]$ to the given input, which can be interpreted as a measure of realism. In contrast, the generator is supposed to transform a random vector y, following a distribution p_Y, into realistic but still synthetic data in order to fool the discriminator. Thus, the generator implicitly defines a probability distribution $p_G = p_{Z|Y}$ which is optimal if it is equal to the actual data distribution p_Z, i.e., if $p_G = p_Z$. Ultimately, the two networks have conflicting objectives and resulting in a zero-sum game that terminates in what is called a Nash-equilibrium. In other words, discriminator and generator play the two-player minimax game

$$\min_G \max_D V(D, G) = \mathbb{E}_{z \sim p_Z}\left[\log D(z)\right] + \mathbb{E}_{y \sim p_Y}\left[\log\left(1 - D(G(y))\right)\right]. \tag{5.26}$$

The desired result is achieved when the generated data is of such realism and the discriminator has only a 50% accuracy, thus resembling random guessing. Although the idea of generative adversarial networks is quite simple, numerous challenges arise during training. Thereby, the training process is mainly based on the following steps:

Step 1: To train the discriminator, both real and synthetic data (sampled from p_G) are needed in equal parts. The real data $z_{\text{real}}^{(i)}$ is labeled with one, while the synthetic data $z_{\text{sim}}^{(i)}$ is labeled with zero.

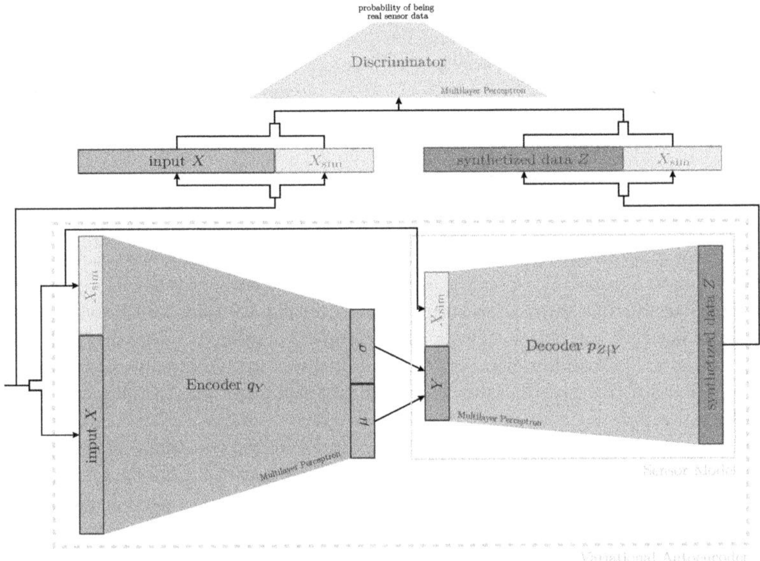

Fig. 5.6. The concept of a generative adversarial network with a variational autoencoder as generator (VAE-GAN). Employing such an architecture, consisting of three independent multilayer perceptrons, removes the necessity of defining a differentiable goodness-of-fit measure. Therefore, the learning process of spatial radar detection distributions might improve.

Step 2: The data sets are fed to the discriminator in terms of mini batches. Subsequently, the weights of the discriminator network are updated by means of backpropagation using the stochastic gradient descent method. It is quite common that steps one and two are repeated several times before the generator is trained in step three. The loss function is given by

$$\frac{1}{n}\sum_{i=1}^{n}\left(\log D(z_{\text{real}}^{(i)}) + \log\left(1 - D(G(y_i))\right)\right). \tag{5.27}$$

Step 3: To train the generator, the weights of the discriminator are fixed. Once again, a mini batch is created by the generator, which consists of purely synthetic data $z_{\text{sim}}^{(i)}$ generated out of random vectors $y_i \sim p_Y$. After evaluation by the discriminator, the gradient is backpropagated through the discriminator in oder to update the weights of the generator only. The error term is given by

$$\frac{1}{n}\sum_{i=1}^{n}\log\left(1 - D(G(y_i))\right). \tag{5.28}$$

As already mentioned, the training successes of generator and discriminator strongly influence each other, therefore one can hardly speak of convergence in a mathematical sense. In general, the training process is extremely unstable [123] and has already been investigated mathematically in more detail [124]. It has turned out in practice that the training often gets unstable when the discriminator outperforms the generator, as the updates to the generator get consistently worse due to a vanishing gradient. Nevertheless, generative adversarial networks yield impressive results because of numerous handy improvements [125], e.g., one-sided label smoothing, feature matching, and, alternative loss functions.

For this very reason, an application to radar point clouds is interesting. However, as with variational autoencoders, such a concept does not guarantee that the synthetic data will match the current simulation state. Therefore, the so-called conditional generative adversarial networks are used. These were already considered in the original paper by Goodfellow et al. [35] and implemented by Mirza et al. [82]. Analogous to the variational autoencoder both the discriminator as well as the generator receive the simulation state as additional input.

A substantial benefit of the adversarial approach is to refrain from defining a differentiable loss to measure the quality of the synthetic data. In fact, for variational autoencoders used to simulate radar detections, this is a major drawback. In particular, the varying number of detections and the spatial association of real and synthetic data often require non-differentiable metrics to evaluate their similarity. Any metrics or quality measures that rely on a nearest neighbor search are undoubtedly non-differentiable.

To circumvent this dilemma, variational autoencoders and generative adversarial networks can be combined to form a VAE-GAN. This was first proposed by Larsen et al. [126] and tested in the case of additional conditions by Bao et al. [127]. Since the discriminator determines the quality of the encoding-decoding process, no additional metrics or loss functions are necessary, thus improving the learning efficiency of spatial structures in particular.

5.4 Comparison of learning capacities and its consequences

In the previous sections, methods for a data based radar point cloud simulation were presented using common techniques. On the one hand a kernel density estimation scheme and on the other hand three methods (variational autoencoder, generative adversarial networks and generative adversarial networks with variational autoencoder as generator) originating from the field of generative modeling using deep neural networks. Based on first experiments, all models are implemented in two stages, determining the number of detections at first and subsequently generating several detections one by one. Finally, the question arises, which of the four methods is objectively best suited to reproduce the spatial distribution of detections, given sufficient data. Since in practice a sufficient, uniformly distributed dataset of high quality cannot be generated in an acceptable time, synthetic data must be used. In order to compare the learning capacities of the different methodologies, several datasets for training, validation and benchmark are generated

Table 5.1. Comparison of different modeling techniques using synthetic data for training and evaluation. Classical statistical approaches based on a kernel density estimation approach are compared with different deep generative networks. Additional benchmark values (BM) are given using a further validation dataset in order to provide an orientation (results already published in [106]).

	BM	VAE	GAN	VAE-GAN	KDE
E_N	1.431	1.866	1.733	1.614	1.773
E_{BB}	0.030	0.097	0.096	0.092	0.056
E_{KLD}	0.008	23.17	19.688	20.714	9.375

using a hybrid modeling approach which will be introduced later on (see chapter 6). The training dataset corresponds to $2 \cdot 10^6$ radar point clouds corresponding to roughly 28 hours of continuous data logging (whereas no uniform distribution of ground truth states would have been achieved after 28 hours). Both the validation and the benchmark dataset correspond to a realistic overtaking maneuver comprising 128 ground truth states and 820 detections (cf. overtaking maneuver in chapter 6). So far, there are no objective quality measures for the evaluation of different radar point clouds. However, since the problem can be traced back to probability distributions, the already mentioned Kullback-Leibler divergence offers a suitable possibility for evaluation. Furthermore, shape parameters are of interest for sensor fusion, as well as for subsequent driving functions. Therefore, an additional measure E_{BB} is defined in order to quantify the deviation of radar point cloud relative to the shape of the surrounding bounding box. For simulated detection lists Z_{sim} and detection lists Z_{valid} of the validation dataset this measure is given by

$$E_{\mathrm{BB}} = \frac{1}{n} \sum_{i=1}^{n} \left\| \mathrm{dist}_{\mathrm{BB}} \left(Z_{\mathrm{real}}^{(i)} \right) - \mathrm{dist}_{\mathrm{BB}} \left(z_{\mathrm{sim}}^{(i)} \right) \right\|_2^2 , \tag{5.29}$$

where $\mathrm{dist}_{\mathrm{BB}}$ represents the mean minimum deviation from the outer edges of the surrounding bounding box. To obtain an overall impression of the peformance, additionally to the Kullback-Leibler divergence and the measure E_{BB}, the number of detections is also of interest. The quality is quantified by E_{N} and represents the mean of the absolute error, i.e.,

$$E_{\mathrm{N}} = \frac{1}{n} \sum_{i=1}^{n} \left| n_{\mathrm{sim}}^{(i)} - n_{\mathrm{valid}}^{(i)} \right| . \tag{5.30}$$

The results of the evaluation are summarized in table 5.1. In general, the smaller the value, the better the performance of the model. Thereby, one can observe a similar performance when modeling the number of detections. Among the three deep generative networks, the variational autoencoder with adversarial loss function (VAE-GAN) reproduces the spatial distribution best. For the Kullback-Leibler divergence, the adversarial modeling approaches are superior to the variational autoencoder. One of the reasons is certainly the lack of a suitable and differentiable loss function when training the variational autoencoder.

In particular, the capabilities of the kernel density estimation method deserve special mention. Compared to deep generative networks, it exhibits significantly lower errors. However, one must admit here that this is not yet a generally valid statement, since not every conceivable network architecture could be tested.
The modeling of radar point clouds based on the kernel density estimation technique has an important advantage, since the method relies on a mathematically rigorous procedure with only few parameters for optimization and furthermore ensures convergence in probability. Although the fundamental theorem deep learning relies on guarantees an approximation of arbitrary functions [128], no convergence statement exists for the training process and will most likely never exist. Furthermore, the training is extremely unstable and the almost endless architectural possibilities for constructing neural networks confront the user with additional challenges whose success is not directly predictable. Therefore, based on the presented evaluation, kernel density estimation methods are currently to be preferred over deep generative networks as radar sensor models.
Nevertheless, data acquisition remains a significant challenge in practice. The modeling was based only on the dimensions of a single vehicle model. In view of the 28 hours of continuous data recording for a single model, generating a data basis with a wide variety of vehicle dimensions seems practically infeasible, especially since reference sensors are constantly required. The training dataset based on a ray-tracing based modeling approach is recommended for further use in standardized simulation environments and should therefore be investigated in more detail (see chapter 6).
Conclusion and recap of the second research question: Statistical methods based on kernel density estimation and neural networks are suitable for a data-based simulation of radar point clouds. However, it must be kept in mind that these methods have a tremendous demand regarding data. Therefore, these approaches are currently considered rather uneconomical. Further drawbacks are the lack of parameterization and the limited applications. However, if one compares the techniques with each other, classical statistical methods are to be preferred over neural networks, contrary to the current trend. The kernel density estimation method yields a reliable training success, can be trained rather easily and results do not require any fine-tuning. Moreover, the training success of kernel density estimation models are even superior to neural networks in significant aspects. In summary one may say that kernel density estimation methods are superior to deep generative methods. Nevertheless, kernel density estimation methods are currently unsuitable for practical use.

Chapter 6

A hybrid modeling approach for radar point clouds

The modeling methods discussed in the previous chapters suffer from some practical shortcomings. Physical ray-tracing methods on the one hand lack material parameters and on the other hand often demand tremendous computational resources to ensure real-time capability. Purely stochastic methods often suffer from a lack of sufficient data and the ability to simulate individual worst-case scenarios in both a fast and specific manner. Moreover, it is unknown what the limitations of the individual training data sets are.

More than ever, these challenges require a hybrid approach to the modeling of automotive radar sensor technology. The term "hybrid" does not refer to a separate modeling approach, but rather combines classical and data-based approaches with the objective of a statistical replication of a particular sensor. For example, simplified physical or ray-tracing based approaches can be used in combination with data-based algorithms in a way that the model reproduces real radar detection distributions as precisely as possible. In contrast to a solely stochastic radar model, a heuristic or physical backup principle is always included to ensure plausible data even in unknown scenarios. Furthermore, a data set of small size should already be sufficient for model adjustment. This is the only way to adapt the model efficiently without requiring more overhead for sensor simulation than for the actual saftey validation.

The following sections present the general, two-dimensional ray casting method followed by a step-by-step extension. Thereby, the approach can always be integrated into a standardized sensor simulation framework (as described in section 1.3 and shown in Fig. 1.6). Initially, the method seems close to the method presented by Knill et al. in [26]. However, additional radar characteristics, such as characteristic scattering centers and micro-Doppler effects, are addressed. Furthermore, a practical application as well as data-based optimizations are discussed.

6.1 Tracing and catching rays as the baseline

At its simplest, the radar sensor is characterized by its field of view, in which one simulates the measurement principle with n rays. If the field of view has an angular range of φ_{FoV},

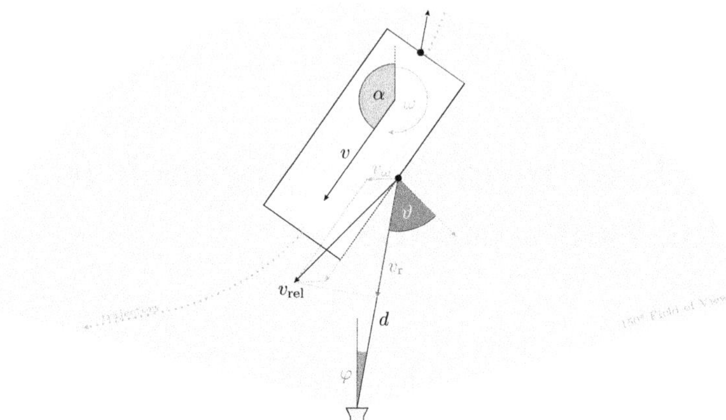

Fig. 6.1. Sketch of the general principles of a two-dimensional ray-tracing radar model and its most important parameters. One should note that up to two detections can occur per ray. Due to the trajectory, the yaw rate ω must also be taken into account to determine the radial velocity v_{r} precisely (cf. [26]).

the rays are emitted equidistantly at a distance of $\Delta\varphi = \varphi_{\text{FoV}}/_{n-1}$. If a ray hits an object, represented by its two-dimensional bounding box and characterized by the state vector

$$\xi = (x, y, v, \alpha, \omega, l, w)\,, \tag{6.1}$$

a radar detection is generated and, if desired, a normally distributed error is added. The state vector, given in the sensor's coordinate system, denotes the position of the rear axis with (x, y), the velocity with v, the yaw angle with α, the yaw rate with ω, and the dimensions of the bounding box with l and w.

At first, this approach is very similar to a simplified lidar sensor model, such as the one described in [31]. However, slight extensions already provide a remedy. In this way, each intersection point initially represents only a potentially possible radar detection. It only becomes an element of the radar point cloud when a random number $u \sim \mathcal{U}[0,1]$ does not exceed the existence probability (sometimes referred to as measurement likelihood) $p(d, \vartheta)$, which is derived from the hit point parameters[1]. On the one hand, the received power decreases with increasing distance d of the object and on the other hand, the angle of incidence determines the direction of reflection. The following two relationships can usually be observed: The more acute the angle or the closer the object, the more detections are generated by the sensor. Furthermore, one can assume that

$$p(d, \vartheta) \approx p_{\text{dist}}(d)p_{\text{ang}}(\vartheta). \tag{6.2}$$

[1]In contrast to [26] the conditional probability considers the angle of incidence directly. An additional velocity influence is omitted due to the principles outlined in section 1.2.4.

Algorithm 1 Two-dimensional ray-tracing radar model

Input: vehicle state $\xi = (x, y, v, \alpha, \omega, l, w)$
Output: radar point cloud $Z = \{z_i\}_{i=1,2,\dots,m}$

$R \leftarrow$ set of emitted rays $\{r_i\}_{i=1,2,\dots,n}$ in direction $\varphi_i \sim \mathcal{N}\left(\frac{\varphi_{\text{FoV}}}{2} - \frac{i-1}{n-1}\varphi_{\text{FoV}}, \sigma^2\right)$
for each r_i **in** R **do**
  if r_i intersects ξ **then**
    $d \leftarrow$ distance to sensor (closest intersection point)
    $\vartheta \leftarrow$ angle of incidence (closest intersection point)
    $u \leftarrow$ random sample from $\mathcal{U}[0, 1]$
    if $u \leqslant p(d, \vartheta)$ **then**
      $k \leftarrow k + 1$ (starting with $k = 0$)
      $v_{\text{r}} \leftarrow$ radial velocity of intersection point
      $\varepsilon \leftarrow$ measurement deviation sample from $\mathcal{E}$
      $z_k \leftarrow (d, \varphi_i, v_{\text{r}}) + \varepsilon$
    end if
  end if
end for
return $Z = \{z_i\}_{i=1,2,\dots,m}$

In particular, this applies to surfaces which are considered smooth according to the Rayleight roughness criterion [129].
At this point, the model generates a highly multimodal detection distribution. What is meant by this is that due to the deterministic beam direction, always the same point is being hit and therefore, despite a measurement deviation, the detections accumulate locally. Since this effect, which cannot be observed in this manner in practice, leads to a highly idealized radar ray casting model further refinements are needed. However, this drawback can be easily circumvented by adding noise to the deterministic ray direction. It might be necessary to ensure that the distance between the detections does not drop below the theoretical minimum.
Another essential point is the parameterization of the existence probabilities. The choice of parameters is significantly related to the number of emitted rays, and thus to the angular density of rays ρ. For the ray density in terms of the distance r one finds

$$\rho(r) = \frac{n-1}{\pi r \frac{\varphi_{\text{FoV}}}{180^\circ}} \sim \frac{1}{r}. \tag{6.3}$$

In other words, on a meter-long circular arc segment with radius r, $\rho(r)$ rays are impinging on average. Given a measurement likelihood that is independent of the object distance, i.e., $p(d, \vartheta) = p_{\text{ang}}(\vartheta)$, the number of hit points would thus automatically decrease with the factor $1/r$. Hence, the minimal number of initially emitted rays can be estimated easily. If an object of width w is to be guaranteed detectable up to a distance r, then

$$\begin{aligned} & w\rho(r) \geqslant 1 \\ \Leftrightarrow \quad & n \geqslant 1 + \frac{\pi r}{w} \cdot \frac{\varphi_{\text{FoV}}}{180^\circ} \end{aligned} \tag{6.4}$$

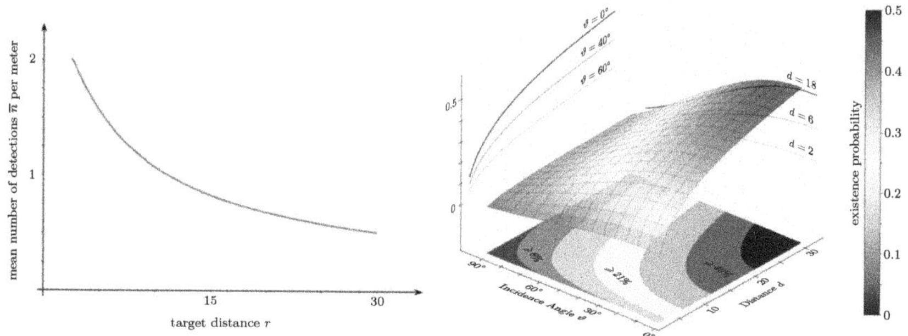

Fig. 6.2. Exemplary parameterization of a ray based model in two dimensions. If the mean number of detections is supposed to decrease as the distance increases (left), similar to a $1/R^4$ law, the existence probability must be monotonically increasing. Without considering the ray density, this fact might seem to contradict a first intuition. The parameterization of the angular dependency is based on simple considerations and follows a cosine window function.

must apply. Consequently, it is not untypical that models consist of one hundred and even more rays. If one does not adjust the existence probability appropriately, too many detections occur in case of short distances. Therefore, the distance-dependent existence probability must be monotonically increasing (see Fig. 6.2). Without considering the ray density ρ, this fact seems to be contrary to the initial consideration that the number of detections decreases when the distance increases. One should note, however, that the measurement likelihood is not directly related to the number of detections. Neglecting the angle, at a distance r the mean number of detections $\overline{n}(r)$ per meter is given by

$$\overline{n}(r) = \rho(r) p_{\text{dist}}(r). \tag{6.5}$$

The parameterization of the angle dependency is much more difficult and cannot be done properly without expertise. Moreover, this parameter also depends on the target object. In general and from a purely heuristic point of view, the angle-dependent measurement likelihood should increase as the angle of incidence becomes more acute. Conversely, large angles may lead to detections in rare cases only, e.g., in the case of complex geometries or numerous edges. Hence, the measurement likelihood tends to zero as the angle of incidence increases. In an exemplary case, one could use an adapted bump function, a generalized cosine window, or any other differentiable window function with a suitable domain. The example given in Fig. 6.2 uses a cosine window, i.e.,

$$p_{\text{ang}}(\vartheta) = \begin{cases} \frac{1}{2}\left(1 - \cos\left(\pi \frac{\vartheta + b}{b}\right)\right) & \text{, for } \vartheta \leqslant b \\ 0 & \text{, for } \vartheta > b \end{cases} \quad \text{and } b \in [0°, 90°]\,. \tag{6.6}$$

Another significant model parameter is the distribution of the measurement deviation. Several models usually assume a normally distributed measurement deviation. Even if this is in fact the case, a normally distributed measurement uncertainty leads to a qualitatively unsatisfactory result, which more or less corresponds to the rectangular contour of the bounding box. Two problems follow from this. On the one hand, the number of detections outside the bounding box is on average equal to the number inside, and on the other hand, the object extent may be overestimated by subsequent algorithms. Moreover, the former phenomenon cannot be observed in practice (see Fig. 6.10). An alternative are continuous, but asymmetric distributions for the deviation of spatial measurements, e.g., a skew-normal distribution, a log-normal distribution, and the Crystal Ball function. In summary, the model presented here – in contrast to the probability-based considerations of Knill et al. – is based on fundamental ideas about radar sensing. Throughout, it attempts to clarify whether a given point reflects the signal emitted by the sensor, and if so, how the resulting detections are distributed. Using the notation of chapter 5, the hit point z satisfies the distribution $p(\cdot \mid d, \vartheta, \xi)$, whereas expert knowledge is consistently required to deduce this distribution.
In order to finally assign the velocity v_r, determined by means of the Doppler effect, to a detection, it is not sufficient to consider the velocity in the driving direction only. Particularly at short distances, e.g., at intersections or traffic circles, yaw rates up to 30° per second might be observed and influence the Doppler effect significantly (see Fig. 6.1). Neglecting the slip, the relative velocity v_rel of the hit point is given as

$$
\begin{aligned}
v_\mathrm{rel} &= v + v_\omega \\
&= \left.\begin{pmatrix} v\cos\alpha \\ v\sin\alpha \\ 0 \end{pmatrix} + \begin{pmatrix} 0 \\ 0 \\ \omega \end{pmatrix} \times \begin{pmatrix} d\cos\varphi - x \\ d\sin\varphi - y \\ 0 \end{pmatrix}\right|_{z=0} \\
&= \begin{pmatrix} v\cos\alpha - \omega(d\sin\varphi - y) \\ v\sin\alpha + \omega(d\cos\varphi - x) \end{pmatrix}.
\end{aligned}
\tag{6.7}
$$

One should note that the notation does not differentiate between velocity of the vehicle as a vector and its magnitude. The velocity v_r, that is measured by means of the Doppler effect, is consequently given by

$$
\begin{aligned}
v_\mathrm{r} &= \begin{pmatrix} \cos\varphi \\ \sin\varphi \end{pmatrix} \cdot \begin{pmatrix} v\cos\alpha - \omega(d\sin\varphi - y) \\ v\sin\alpha + \omega(d\cos\varphi - x) \end{pmatrix} \\
&= v\cos(\varphi - \alpha) + \omega(y\cos\varphi - x\sin\varphi).
\end{aligned}
\tag{6.8}
$$

However, it should be noted that the size of the bounding box does not correspond to the actual dimensions of the vehicle and, therefore, the angular velocity v_ω might be over- or underestimated. Accordingly, the radial velocity v_r also deviates from the actual radial velocity. In general, though, these deviations can be neglected. In early development stages, or when there is no suitable database, the measurement uncertainty is preferably simulated with a Gaussian white noise based on the parameters provided by the specifications given in the sensors data sheet, e.g., for the short range radar sensor SRR520 provided by Continental $\sigma^2 = 0.1\,\mathrm{km\,h^{-1}}$ [130].

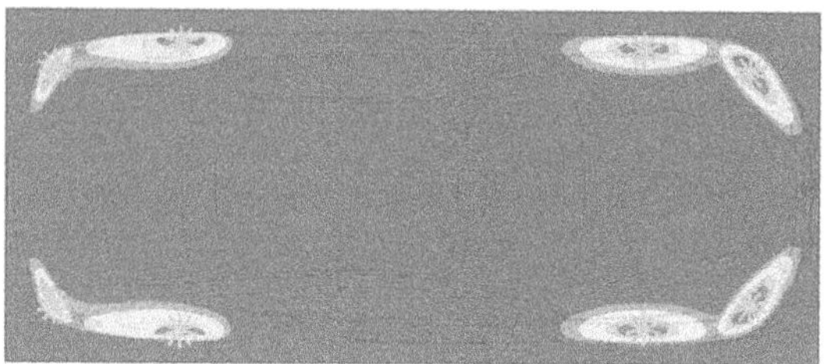

Fig. 6.3. Characteristic scattering centers and their measurement deviation distribution. Abstract distribution functions (red contour lines) are used to simulate the detection distribution of a given sensor in great detail. Even though the ground truth description is based on rather simple abstract shapes.

6.2 Improvements to the ray casting approach

The model presented in the previous section is easy to implement, real-time capable and can be parameterized quickly. If one looks at real detection distributions (see Fig. 6.10), the model nevertheless has a decisive shortcoming. The reason is that the underlying distributions cannot be simulated solely on the basis of a bounding box. Therefore, additional parameters are necessary for the detailed simulation of vehicular contours.
In the following sections, the model is extended to include characteristic scattering centers as first presented by Bühren et al. in [62]. This extension is also perfectly suited for modeling wheel speed effects, such as the so-called micro-Doppler effect , and reflections on wheels that are only partially visible (see Fig. 6.4). Of particular interest are the theoretically measurable radial velocities, which will be discussed in detail in the second subsection. Finally, further ideas on data-based model fitting are presented and the benefits and drawbacks of the model are highlighted.

6.2.1 Characteristic scattering centers

If one looks at accumulated real data of radar sensors, one discovers significant accumulations at certain locations. These accumulations have already been discussed in detail by numerous authors and have been considered, for example, by Bühren et al. in the modeling of automotive radar sensor technology. Therefore, a consideration of those characteristic scattering centers is also desirable in the case of ray casting and -tracing models. In the following, this approach is adopted and integrated into the model based on eight backscattering centers. Although headlights, tail lights and wheel housings including rims and tires are explicitly considered, almost any number of points can be realized. In the limiting case, i.e., when the number of scattering centers tends to infinity, distribution functions for the backscattering properties are modeled in this way.
Up to now, the hit point position has not been taken into account in the two-dimensional ray casting model. Neither in the modeling measurement deviation, nor in the existence

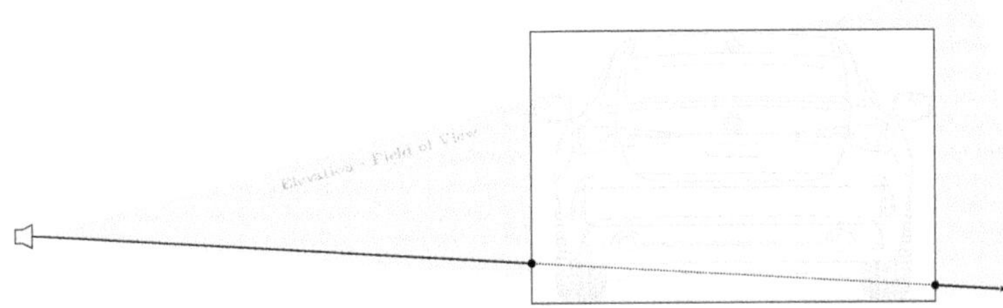

Fig. 6.4. Illustration of wheel detections on the opposing side. Due to the underbody clearance, it is possible that the sensor may also detect the wheels on the right side. The extended model can take this into account if the exit point is close to a wheel, i.e., close to a certain characteristic points.

probability. However, since such detections occur considerably more often in practice, their existence probability must also differ significantly. Therefore, one should distinguish between the measurement likelihood of ordinary detections, i.e.,

$$p_{\text{ord}}(d, \vartheta) = p_{\text{dist}}^{(0)}(d) p_{\text{ang}}^{(0)}(\vartheta), \tag{6.9}$$

and detections at n_{char} characteristic scattering centers, i.e.,

$$p_{\text{char}}^{(i)}(d, \varphi, \vartheta) = p_{\text{pos}}^{(i)}(d, \varphi, \vartheta) p_{\text{dist}}^{(i)}(d) p_{\text{ang}}^{(i)}(\vartheta) \quad \text{for } i = 1, 2, \ldots, n_{\text{char}}. \tag{6.10}$$

Thereby, $p_{\text{pos}}^{(i)}$ can be used to take the point of impact into account, i.e., front, back, left, or right. This can be particularly relevant in the case of headlights, for example. If the sensor is located behind the headlight, and thus the rays hit the bounding box on the left side, its reflection properties differ significantly from those of a corner reflector. Although in the preliminary considerations it was assumed that characteristic detections occur more frequently and hence $p_{\text{char}}^{(i)}(d, \vartheta) \geqslant p_{\text{ord}}(d, \vartheta)$ would hold, it is reasonable to deviate from this assumption in situations like this.

Now, whether a hit point can be assigned to a characteristic scattering center or not can be decided depending on the distance (these ranges can and should also differ from point to point) and the underlying measurement deviation distribution $\mathcal{E}_{\text{char}}^{(i)}$. In connection with the characteristic points, no restrictions should be specified for the measurement deviations, since this allows a visually extraordinarily detailed emulation of the detection distributions. Furthermore, despite the high abstraction of a standardized ground truth description, e.g., the Open Simulation Interface, vehicle contours can also be exactly reproduced in the detection image (see Fig. 6.3).

Another significant aspect of ray casting in combination with characteristic points is the consideration of the underbody clearance. If the sensor installation position is sufficiently low or the distance to the target vehicle is large enough, the wheels can be detected in numerous situations due to the underbody clearance, even though the side of the vehicle is not directly hit by rays. This effect is described in Fig. 6.4 and has already been

investigated by Kellner et al. [131]. However, due to the rotation of the wheels, additional challenges arise for the simulation of the velocity profile (see subsection 6.2.2).
In order to take all these effects into account, the vehicle state ξ must be extended. The state must include at least the vehicle model t_{type}, e.g., BMW 740i (G12), and the steering angle δ. Information about the characteristic points, their distance tolerances, distributions and measurement likelihoods are preferably provided by a look-up table. Hence, the state of the target vehicle is described by

$$\xi = (x, y, v, \alpha, \delta, \omega, l, w, t_{\text{type}}) \,. \tag{6.11}$$

However, it must be mentioned that in the current version of the Open Simulation Interface (version 3.2.0) the steering angle δ is not included and possibly not considered by vendors of driving simultations. In addition, the vehicle model may also not be taken into account [132]. Although it seems desirable to take both parameters into account, it is not compulsory to do so. For example, the steering angle can be derived from the single-track model using the yaw rate ω, the velocity v and the object dimensions [133]. The vehicle model can be provided – if the environment simulation provides it – via optional information fields. A complete pseudecode of the extended two-dimensional ray-tracing radar model is given in Algorithm 2.

6.2.2 The Doppler effect of wheels and tires

As already repeated in detail at the beginning of the thesis, the Doppler effect allows the measurement of the target's velocity. On the one hand the electromagnetic wave emitted by the sensor is compressed or stretched – depending on the direction of motion of the target – and on the other hand a relativistic time transformation takes place. Since a target vehicle (and of course also pedestrians, cyclists, etc.) is not a purely rigid body, various Doppler shifts occur due to additional motions. Therefore, objects are not only exhibiting a range expansion, but also an extended velocity profile (see Fig. 6.5).
Consequently, additional detections occur due to the motion of wheels. Since this information is already used by algorithms for object detection and object tracking, these effects must be simulated as faithfully as possible, too. With the help of characteristic scattering centers, this phenomenon can be simulated easily. In case a ray of the ray casting model hits the target close to a wheel (the position depends on the vehicle type, but in the simplest case it can be given as a percentage of the object's extent), the location of the detection is sampled from the distribution belonging to that scattering center. In contrast to all other detections, however, strongly fluctuating radial velocities can be measured here. How these radial velocities are distributed in principle will be discussed in the following.
Assuming the target vehicle is again described by the state vector ξ and the ray casting model hits the center of the wheel, its velocity is given according to the equation (6.7) by

$$v_{\text{rel}} = \begin{pmatrix} v \cos\alpha - \omega(d \sin\varphi - y) \\ v \sin\alpha + \omega(d \cos\varphi - x) \end{pmatrix} \tag{6.12}$$

and points in direction of the wheel position, i.e., v_{rel} and v are positioned at the steering angle δ to each other. However, if the center of the wheel itself is not directly detected, there is an additional velocity influence due to the rotational motion of the wheel.

Algorithm 2 Extended two-dimensional ray-tracing radar model

Input: vehicle state $\xi = (x, y, v, \alpha, \delta, \omega, l, w, t_{\text{type}})$
Output: radar point cloud $Z = \{z_i\}_{i=1,2,\dots,m}$

$R \leftarrow$ set of emitted rays $\{r_i\}_{i=1,2,\dots,n}$ in direction $\varphi_i \sim \mathcal{N}\left(\frac{\varphi_{\text{FoV}}}{2} - \frac{i-1}{n-1}\varphi_{\text{FoV}}, \sigma^2\right)$
for each r_i **in** R **do**
  if r_i intersects ξ **then**
    $d_1, d_2 \leftarrow$ distances to sensor $(d_1 \leqslant d_2)$
    $\vartheta_1, \vartheta_2 \leftarrow$ angles of incidence
    $l_1, l_2 \leftarrow$ location of the hit points {front, rear, left, right}
    $u_1, u_2 \leftarrow$ random samples from $\mathcal{U}[0, 1]$
    # Simulation of directly visible hit point, i.e., the closest one:
    if (d_1, φ_1) close to a scattering center c_i **and** $u_1 \leqslant p_{\text{char}}^{(i)}(d_1, \vartheta_1, \xi)$ **then**
      $v_{\text{r}} \leftarrow$ radial velocity of intersection point
      $\varepsilon \leftarrow$ measurement deviation sample from $\mathcal{E}_{\text{char}}^{(i)}$
      $Z \leftarrow$ add $(c_i, v_{\text{rel}}) + \varepsilon$ to radar point cloud
    else if $u_1 \leqslant p_{\text{ord}}^{(l_1)}(d_1, \vartheta_1, \xi)$ **then**
      $v_{\text{r}} \leftarrow$ radial velocity of intersection point
      $\varepsilon \leftarrow$ measurement deviation sample from $\mathcal{E}_{\text{ord}}^{(l_1)}$
      $Z \leftarrow$ add $(d, \varphi_i, v_{\text{rel}}) + \varepsilon$ to radar point cloud
    end if
    # Simulation of the second hit point, i.e., the most distant one:
    if (d_2, φ_2) close to a scattering center c_i **and** $u_2 \leqslant p_{\text{char}}^{(i)}(d_2, \vartheta_2, \xi)$ **then**
      $v_{\text{r}} \leftarrow$ radial velocity of intersection point
      $\varepsilon \leftarrow$ measurement deviation sample from $\mathcal{E}_{\text{char}}^{(i)}$
      $Z \leftarrow$ add $(c_i, v_{\text{rel}}) + \varepsilon$ to radar point cloud
    end if
  end if
end for
return $Z = \{z_i\}_{i=1,2,\dots,m}$

Kellner et al. have pointed out in [131] that, for instance, the contact point of the tire and the road has no radial velocity component (assuming a yaw rate $\omega = 0$; in the coordinate system of the sensor), but the opposing side of the tire (with respect to the center of the wheel) shows twice the relative velocity. In order to examine the distribution of the velocity components, which obviously appear to be within the interval $[0, 2v_{\text{r}}]$, the relative velocity v_{rel} at the center of the wheel and the radius r_{w} of the wheel are given below. Thus, the angular velocity ω_{w} of the wheel is given by

$$\omega_{\text{w}} = \frac{\|v_{\text{rel}}\|}{r_{\text{w}}}. \tag{6.13}$$

At an arbitrary point $(x_{\text{w}}, 0, z_{\text{w}})$ on the visible side of the wheel (where the x-axis of the right-handed coordinate system points in direction of motion and parallel to the ground, see Fig. 6.6), the absolute value of the velocity due to the rotational motion is therefore

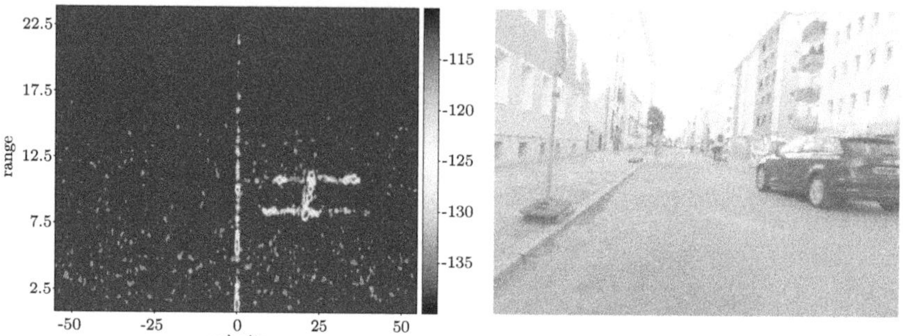

Fig. 6.5. Measurement-based illustration of the micro-Doppler effect. Within the range-velocity-map the target vehicle (see camera view) has different velocity components. Consequently, the velocity measurement at wheels differ in parts significantly compared to the vehicle velocity [134].

given by

$$\|v_{\text{rot}}\| = \omega_{\text{w}}\sqrt{x_{\text{w}}^2 + z_{\text{w}}^2}. \tag{6.14}$$

The velocity vector associated to the point $(x_{\text{w}}, 0, z_{\text{w}})$ lies inside the xz-plane and is perpendicular to the position vector. Therefore

$$v_{\omega_{\text{w}}} = \omega_{\text{w}} \begin{pmatrix} z_{\text{w}} \\ 0 \\ -x_{\text{w}} \end{pmatrix} \tag{6.15}$$

holds. In the three-dimensional coordinate system of the sensor, both steering angle δ and yaw angle α must be taken into account. In order to investigate the influence on the relative velocity, it suffices to consider its two-dimensional projection. Hence, only the vector

$$\begin{aligned} v_{\omega_{\text{w}}}^{\text{w}} &= \omega_{\text{w}} \left.\begin{pmatrix} z_{\text{w}}\cos(\alpha+\delta) \\ z_{\text{w}}\sin(\alpha+\delta) \\ -x \end{pmatrix}\right|_{z=0} \\ &= z_{\text{w}}\omega_{\text{w}} \begin{pmatrix} \cos(\alpha+\delta) \\ \sin(\alpha+\delta) \end{pmatrix} \end{aligned} \tag{6.16}$$

has to be considered in the following. Therefore, the additional velocity contribution in radial direction to the sensor only depends on the z-component of the point. On top of the motional speed of the entire wheel, there is an additional contribution of $v_{\omega_{\text{w}}}^{\text{w}}$, i.e., the actual relative velocity $v_{\text{rel,rot}}$ of a point on the wheel is given by

$$v_{\text{rel,rot}} = v_{\text{rel}} + v_{\omega_{\text{w}}}^{\text{w}}. \tag{6.17}$$

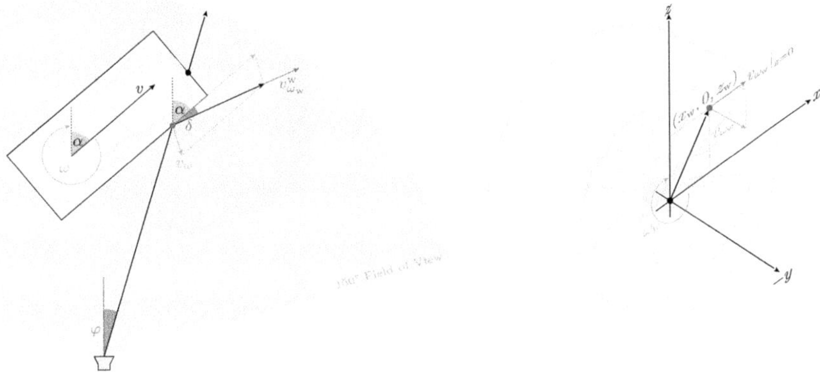

Fig. 6.6. Sketch for the derivation of the additional detectable velocity components due to the rotational motion of the wheel. On the left side a representation according to the sensor model perspective and on the right side a detailed representation in the coordinate system of the wheel.

In analogy to equation (6.8), the detectable velocity due to the Doppler effect is given by the projection onto the normalized ray vector of the model, i.e.,

$$\begin{aligned} v_{\mathrm{r,rot}} &= \begin{pmatrix} \cos\varphi \\ \sin\varphi \end{pmatrix} \cdot \left(\begin{pmatrix} v\cos\alpha - \omega(d\sin\varphi - y) \\ v\sin\alpha + \omega(d\cos\varphi - x) \end{pmatrix} + z_{\mathrm{w}}\omega_{\mathrm{w}} \begin{pmatrix} \cos(\alpha+\delta) \\ \sin(\alpha+\delta) \end{pmatrix} \right) \\ &= v\cos(\varphi-\alpha) + \omega(y\cos\varphi - x\sin\varphi) + z_{\mathrm{w}}\omega_{\mathrm{w}}\cos(\varphi-\alpha-\delta). \end{aligned} \tag{6.18}$$

Now, one easily recognizes the previously mentioned relationship that for a negligible yaw rate $\omega \approx 0$, respective for negligible steering angle $\delta \approx 0$, at the two points $(0, 0, \pm r_{\mathrm{w}})$

$$\begin{aligned} v_{\mathrm{r,rot}} &= v\cos(\varphi-\alpha) \pm \omega_{\mathrm{w}} r_{\mathrm{w}} \cos(\varphi-\alpha) \\ &= v\cos(\varphi-\alpha) \pm v\cos(\varphi-\alpha) \\ &= (v \pm v)\cos(\varphi-\alpha) \end{aligned} \tag{6.19}$$

holds. Furthermore, in principle, all other velocities in the interval $[0, 2v\cos(\alpha-\varphi)]$ can be encountered. If one looks carefully at Fig. 6.7, one can see that, in theory, there are more detectable points exhibiting a velocity of v than, for instance, with velocity $0.5v$ or $2v$. This is due to the simple fact that the wheel area visible to the radar sensor is at its largest near $z = 0$ and decreases steadily as the absolute value of z increases. Considering only the rim and neglecting the tire, the distribution for the theoretically detectable velocities in case of an idealized solid rim can be determined as follows. The yaw angle of the wheel is $\alpha + \delta$, thus the actually perfectly circular rim appears to be approximately elliptical when projected in two dimensions. In other words, the area visible to the sensor, i.e., the area penetrated by radar waves, is elliptical, i.e., in the horizontal direction the extension of the wheel is only $2r_{\mathrm{w}}\sin(\alpha+\delta)$, while the vertical

Fig. 6.7. Quantitative representation of different velocity components as seen by the sensor. On the left the angular velocities of the wheel and on the right the detectable velocities. Blue corresponds to a velocity of $v = 0$ and red to $v \approx 1.7v$ (see also Fig. 6.8 for a qualitative presentation).

extension remains unchanged at $2r_\mathrm{w}$. The visible range in the yz-plane of the sensor is consequently bounded by

$$y = \pm\sqrt{r_\mathrm{w}^2 - z_\mathrm{w}^2}\,\sin(\alpha + \delta). \tag{6.20}$$

Now, the velocity distribution can be derived from this equation, whereby r_w corresponds to the maximum detectable velocity $v\cos(\varphi - \alpha)$. If one transforms equation (6.20) into the velocity domain, one obtains the probability density function f_{ω_w} of the rotation-based and visible velocity components of an idealized solid rim, i.e., with an integration constant $c(\varphi, v, \alpha) > 0$

$$f_{\omega_\mathrm{w}}(x) = c(\varphi, v, \alpha)\sqrt{1 - \left(\frac{y}{v\cos(\varphi - \alpha)}\right)^2} \quad \text{for } x \in [-1, +1]\, v\cos(\varphi - \alpha) \tag{6.21}$$

applies. If the initial speed of the target vehicle is taken into account, i.e., added on top, the probability density function $f_{\mathrm{r,rot}}$ of the wheel speeds $v_{\mathrm{r,rot}}$ detectable by the sensor is subsequently given by

$$f_{\mathrm{r,rot}}(x) = c(\varphi, v, \alpha)\sqrt{1 - \left(\frac{x}{v\cos(\varphi - \alpha)} - 1\right)^2} \quad \text{for } x \in [0, 2v\cos(\varphi - \alpha)]\,. \tag{6.22}$$

Finally, the area, and hence the measurement probability $P(v_{\mathrm{r,rot}} \in I)$ for a given velocity interval $I = [v_1, v_2]$, with $0 \leqslant v_1 < v_2 \leqslant v_{\mathrm{r,rot}}$, can be determined by integration, i.e.,

$$P(v_{\mathrm{r,rot}} \in I) = \int_{v_1}^{v_2} f_{\mathrm{r,rot}}(x)\mathrm{d}x. \tag{6.23}$$

For example, if the wheel is directly in front of the sensor and is therefore hit by a ray with $\varphi = 0°$, then in case of a yaw angle of $\alpha = 25°$, the probability for a velocity in the range of $[1.7v, 2v]$ is about 9.4% (this result was already presented in [135]).

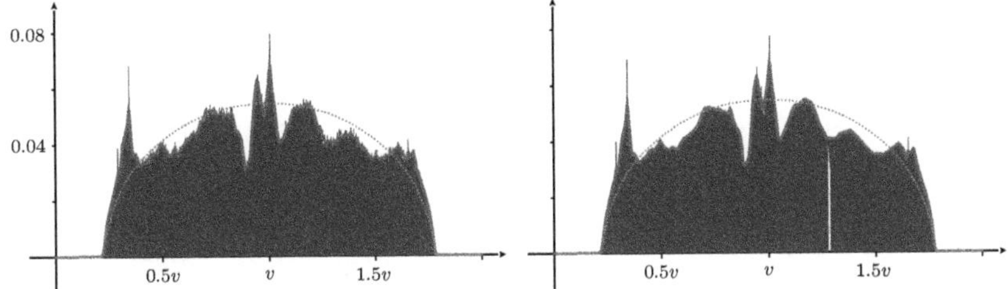

Fig. 6.8. Visualization of the probability density function for the velocity profile of a real rim (cf. Fig. 6.7) determined using a separate ray casting approach. On the left the density at a specific moment and on the right during an entire sensor measurement cycle. The dashed line represents a solid rim

It is important to note, though, that this example only takes the visual properties of a theoretically solid rim and a tire with the same, perfect reflective properties into account. Obviously, these characteristics are not met by a real rim. If one assumes that the tire does not reflect radar beams (this assumption is not necessarily reasonable due to the tire's steel belt), then depending on the sidewall height h of the tire, only velocities in the interval $[{}^{h}/_{r_\mathrm{w}}, 2 - {}^{h}/_{r_\mathrm{w}}]\, v \cos(\varphi - \alpha)$ are discoverable. Due to the abstract geometry of real rims, the probability density function cannot be determined analytically anymore. Nevertheless one can use a separate ray casting method for approximation.
What a suitable probability density function of the rim shown in Fig. 6.7 looks like can be found in Fig. 6.8. Using a separate ray casting approach, the visible regions of the rim were sampled several million times. For each hit point, the rotation based velocity is subsequently determined depending on the distance to the center of the wheel. As previously shown, though, only the z-component of the point has an effect on the speed detectable by the sensor (shown on the right in Fig. 6.7). If one determines the relative frequencies of these velocities, one obtains an approximation of the actual density function (cf. Fig. 6.8). It should be noted, by the way, that the wheel rotates during a sensor measurement cycle and the density function may change (slightly). This seems negligible at first for typical measurement periods of a radar sensor, but the following calculation shows that this is not necessarily the case. Assuming, as an example, a vehicle speed of $v = 15\,\mathrm{m\,s^{-1}}$ and 245/45 R19 tires, during a measurement period of $t_\mathrm{m} = 5.12\,\mathrm{ms}$ (which corresponds to 256 chirps á 20 µs) there is an angular change of

$$t_\mathrm{m} \cdot \frac{v}{\pi r_\mathrm{rim}} \cdot 360^\circ \approx 12.5^\circ. \tag{6.24}$$

Therefore, the densities that arise throughout the entire measurement cycle should be averaged (see right side of Fig. 6.8). Subsequently, the averaged probability density can be used for the simulation of the micro-Doppler effect together with the already presented (idealized) ray casting model. Then, if a ray hits the wheel, i.e., a characteristic scattering center, the velocity measurement is simulated through sampling from the distribution just

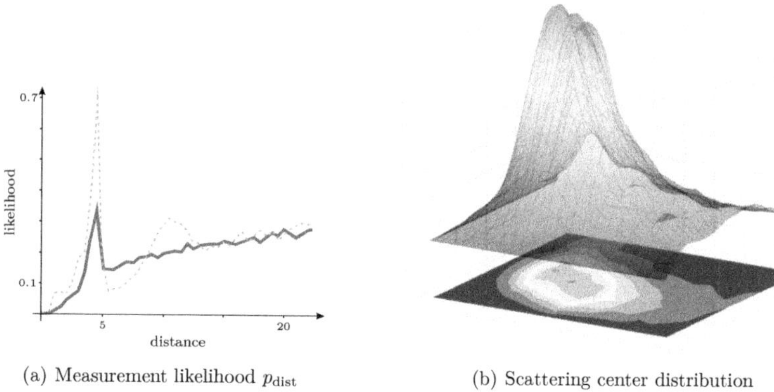

(a) Measurement likelihood p_{dist} (b) Scattering center distribution

Fig. 6.9. Illustration of data-based optimization possibilities. On the one hand, the distance-dependent measurement likelihood can be adapted to match real characteristics (see (a) whereby the dashed line represents real data and the blue line represents model data). On the other hand, a manual labeling of real data together with a two-dimensional curve-fitting can be used to emulate the detection distribution of characteristic scattering centers (see (b)).

derived or determined with ray casting methods.

During the derivation of the probability density functions, it was implicitly assumed that all visible points of the rim are capable of reflecting the electromagnetic wave back to the transmitter. In most cases, of course, this is not the case. Instead, the visible points and their velocity should be weighted by a factor when deriving the probability density function. One way to determine this factor is with the help of a computer aided calculation of the radar cross section. If this is also based on a ray-tracing method, the contribution to the overall radar cross section can be ascertained for each point of the rim. Afterwards, its contribution can be used as a weighting factor when calculating the relative frequencies. Currently, this is offered by the commercial tool CST Studio Suite, and most likely by others as well, but only quantitatively and visually (named hot-spot visualization), thus making precise weighting not possible at this time [136]. Finally, the relationship to Fig. 6.5 can be described by examining the actual velocity components within the range-velocity-map on the basis of an precise radar simulation as done by Wald and Weinmann in [56].

6.3 Capabilities for data-based optimization

The presented approach allows a comprehensive and detailed simulation of the spatial radar detection distribution. Nevertheless, each sensor behaves differently, thus requiring a re-parameterization for each sensor. Although radar point clouds already attain a considerable level of detail in this way, data-based optimization is essential in order to virtualize a given sensor as precisely as possible. Especially within a standardized sensor simulation framework and in the context of virtual safety validation of advanced driver assistance systems. In the following, several methods of data-based optimization will be presented.

6.3.1 Adapting the distance-dependent measurement likelihood

In Fig. 6.2 an exemplary parameterization of the existence probabilities has already been performed. In order to derive a data-based distance-dependency in the following, the general parameterization procedure is briefly described once again. Based on the data sheet of the sensor, a measurement likelihood for the near and far range can be derived. Assuming that the target vehicle is detected at a certain distance with a high probability, at least one detection has to be generated on average. Conversely, the maximum number of near range detections (which in turn is also related to the angular resolution of the sensor) determines the measurement likelihood in close range. For the number of detections, one might assume either a linear or an exponential (as in Fig. 6.2) trend and thus determine the probability function. Nevertheless, this assumption does not necessarily match to the behavior of the given sensor.

To circumvent this issue, real data with additional ground truth (e.g., generated by a differential global positioning system combined with an inertial navigation system) can be used. If possible, the data consists of nearly identical scenarios, so that an average number of detections n can be determined for each timestamp, i. e. for each vehicle state. Subsequently, these trajectories are re-simulated based on an (ideal) ray casting model with a distance-dependent measurement likelihood $p_{\text{dist}}(d) = 1$. Thereby, a sufficient ray density must be ensured.

If the angular range $\Delta\varphi$ and the mean distance d_{m} of all detections are determined, the length w of the arc segment penetrated by the rays can be determined. Accordingly, one finds

$$w = \pi d_{\text{m}} \frac{\Delta\varphi}{180^\circ}. \tag{6.25}$$

For the measurement likelihood, analogous to equation (6.5), the following applies in an approximate manner

$$p_{\text{dist}}(d) \approx \frac{n}{w\rho\left(d_{\text{m}}\right)}. \tag{6.26}$$

Performing this estimation for as many different distances of the target vehicle as possible, one obtains a distance-dependent measurement likelihood (see Fig. 6.9(a)). If the initial ray density is too low, probability values greater than one may be encountered. In this case, the ray density of the model may need to be adjusted. Since this method neglects the angle of incidence, one should be careful that the data set covers mainly highway scenarios and similar maneuvers.

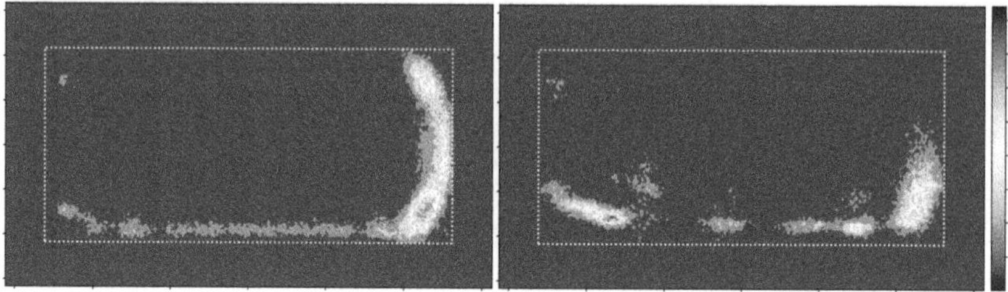

Fig. 6.10. A comparison of synthetic data from the hybrid radar model based on an idealized model parameterization (left) and real sensor data (right) during numerous, identical overtaking maneuvers. The front of the vehicle is positioned towards the right in both cases. The example reveals a non-optimal parameterization of the angle dependency (see left front headlight).

6.3.2 The spatial measurement deviation of scattering centers

The definition of characteristic scattering centers can be used to assign vehicle contours and model properties to the distribution of radar detections even in case of standardized simulation frameworks and an abstract object representation. In the early stages of development, however, it might be useful to work with simple distributions. As an example, the detection distribution on headlights can be simulated with a two-dimensional normal distribution. For wheel cases a combination of symmetric and skew symmetric normal distribution (cf. left side of Fig. 6.10) fits well from a visual point of view. Unfortunately, this method does not necessarily virtualize a particular radar sensor (the data of a simply parameterized model and of a real sensor can be found in Fig. 6.10).

Additional ground truth (e.g., generated by a differential global positioning system combined with an inertial navigation system) and real sensor data of various driving maneuvers can be used to determine the actual distributions significantly better. A priori, it is not obvious which detections should be assigned to a characteristic scattering center. Although, for example, the use of an expectation–maximization algorithm [107, p. 424 ff.] for an unsupervised optimization of the distribution parameters is conceivable, initial experiments have not yielded satisfactory results. Moreover, enormous runtime is often required for simultaneous parameter optimization.

Instead, a manual labeling of the real data together with a two-dimensional curve-fitting allows a fast and precise replication (see Fig. 6.9(b)). Subsequently, these distributions can be integrated into the model. In addition, the manual labeling provides an additional option to automatically determine the probability density function for the velocity measurement at wheels.

6.4 Bottom line on the hybrid modeling approach

The hybrid modeling approach presented in this chapter combines different approaches to radar point cloud simulation and can be deployed in a standardized sensor simulation framework. No similar approaches exist so far and in contrast to purely data-driven methods, this approach can already be used in very early development stages. As development progresses and real data becomes available, the model can be refined and adapted to the real sensor in a continuous manner. Furthermore, even if certain scenarios are not covered by the training data, the ray based heuristic ensures a reliable simulation. Consequently, the model can be used in manifold ways.

Nevertheless, it must be noted that the model does not incorporate numerous radar-specific effects (such as those cited by Holder et al. in [22]). In principle, though, the model can easily be extended to include occlusions and multipath propagation using a ray-tracing approach. This requires detailed investigations on the choice of parameters. A further conceivable extension is the consideration of antenna patterns by means of an additional factor for the measurement likelihood in dependence of the ray angle.

In summary, hybrid modeling approaches based on ray-tracing and casting provide an ideal opportunity to virtualize individual sensors within standardized sensor simulation frameworks. In contrast to purely data-based models, the necessary database can be obtained easily. However, additional radar effects remain a subject of research.

Conclusion and recap of the third research question: There are indeed appropriate approaches to the simulation of radar point clouds. Also a combination of data-based and simplified physical concepts is possible, whereby the model remains parametrizable and extendable. Based on the presented model, large areas of development and testing can be covered.

Chapter 7

Validation based on statistical hypothesis testing

The validation of radar sensor models is currently facing substantial ongoing challenges, and their solutions are becoming more and more urgent due to the progress in development. Although the validation concept of Holder et al. [91] provides a thoroughly useful approach, certain challenges remain unsolved. This is similarly true for related areas, e.g., lidar sensor model validation. Therefore, in the next section, an additional criterion for sensor model validation is defined to increase the acceptance of future methods. Afterwards, the mathematical foundations for a validation method based on statistical hypothesis testing are presented and subsequently extended to radar sensor model validation. After a first examination, an outlook on this exciting topic is given.

7.1 Consistency of validation criterion

Like already mentioned in section 2.4, acceptance, intuitiveness and interpretability of validation methodologies is of utmost importance. However, it is the thoughtless application of metrics in combination with a threshold-based validation criterion that compromises this. The following example is intended to illustrate this once again (similarly illustrated in a previous publication [135]) and deals solely with the evaluation of the spatial distribution of detections.

In a static scenario, the radar detections are logged during a single measurement cycle. The target is an ordinary vehicle (see Fig. 7.1). In the Cartesian coordinate system of the sensor, detections occur at

$$Z_{\text{real}} = \left\{ \begin{pmatrix} 2.5 \\ 1.55 \end{pmatrix}, \begin{pmatrix} 2.2 \\ 2 \end{pmatrix}, \begin{pmatrix} 2.25 \\ 2.61 \end{pmatrix}, \begin{pmatrix} 2.95 \\ 1.5 \end{pmatrix}, \begin{pmatrix} 4.05 \\ 0.75 \end{pmatrix}, \begin{pmatrix} 5.25 \\ 0.25 \end{pmatrix} \right\}. \tag{7.1}$$

As suggested by most experts, the static scenario is subsequently re-simulated. Without loss of generality, synthetic radar detections are generated with the help of an illustrative sensor model. Since even the sensor does not provide identical detections when measured once again, it is not surprising if the model generates detections at entirely different locations, i.e., (again in the Cartesian coordinate system of the sensor)

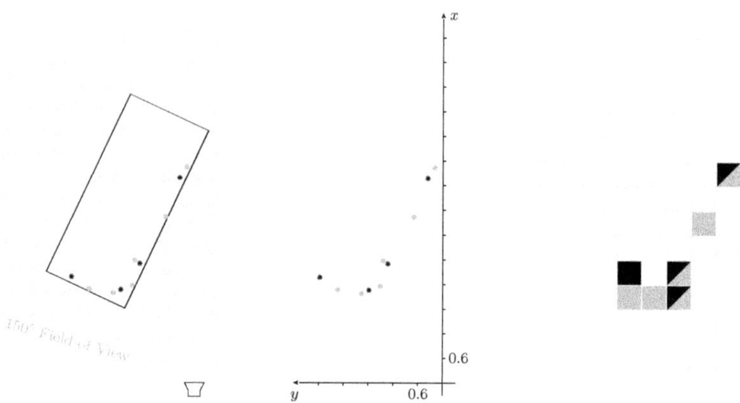

Fig. 7.1. Illustration of the numerical example. On the left hand side the two-dimensional top view, in the center there are the same detections in the sensor coordinate system, and on the right hand side the representation based on an occupancy grid map. The real detections are shown as gray dots, the re-simulated ones as black stars. The question arises: How valid is the model?

$$Z_{\text{sim}} = \left\{ \begin{pmatrix} 2.25 \\ 1.8 \end{pmatrix}, \begin{pmatrix} 3 \\ 2.5 \end{pmatrix}, \begin{pmatrix} 2.9 \\ 1.4 \end{pmatrix}, \begin{pmatrix} 5 \\ 0.4 \end{pmatrix} \right\}. \tag{7.2}$$

So now the question arises whether the model output is within an accuracy range appropriate for the intended application of the model. For point cloud based lidar sensor models, Schaermann [89] and Hanke [31] suggest, for example, the overall error considering the two occupancy grid maps G_{real} and G_{sim} (see Fig. 7.1). In this example (and with a coarse fictitious rasterization) the overall error is calculated as

$$\sum_{i,j} |G_{\text{real}}(i,j) - G_{\text{sim}}(i,j)| = 4. \tag{7.3}$$

Although in this case the result is still rather simple to interpret (number of non-matching cells), the experiments of the two authors show that this is no longer the case in practice due to the large numbers involved for fine granular occupancy grids. The subsequent evaluation of the model validity is thus, on the one hand, lacking the often cited intuitiveness and on the other hand, due to the severe noise variation of real radar data, it cannot be excluded that even a second measurement will satisfy such a simple threshold criterion.

In order to make validation methods in the area of radar sensor modeling as intuitive and interpretable as possible for future use, a mathematical formulation is advantageous. Consequently, a validation should be considered as a mapping

$$V\colon \mathcal{Z}_{\text{real}} \times \mathcal{Z}_{\text{sim}} \to [0,1]. \tag{7.4}$$

Thereby $\mathcal{Z}_{\text{real}}$ and $\mathcal{Z}_{\text{sim}}$ represent the set of all possible detection lists (in the previous example the power set of $\mathbb{R}^2$ without intervals). The validation function V takes real and synthetic detections and returns values between zero and one. Consequently, the result of the validation process can be understood as the degree of validity.
As mentioned in the previous example, due to the large uncertainties it is quite conceivable that the two data sets Z_{real} and Z_{sim} were both recorded with a real sensor, i.e., $Z^{(1)}_{\text{real}}$ and $Z^{(2)}_{\text{real}}$. If one applies the validation function V to these two sets of data, a positive result shall be returned. Obviously, the sensor itself is a perfect model on its own. Hence it is mandatory that

$$V\left(Z^{(1)}_{\text{real}}, Z^{(2)}_{\text{real}}\right) \approx 1 \tag{7.5}$$

holds. Although the optimal value of one is likely to be unattainable in most applications. Conversely, of course, for two synthetic datasets $Z^{(1)}_{\text{sim}}$ and $Z^{(2)}_{\text{sim}}$

$$V\left(Z^{(1)}_{\text{sim}}, Z^{(2)}_{\text{sim}}\right) \approx 1 \tag{7.6}$$

should also apply. Despite the fact that these conditions seem to be quite natural, no such requirements for validation have been defined in the area of sensor modeling so far, nor has it been verified whether previous validation approaches are consistent in this regard. Therefore, in the following the equations (7.5) and (7.6) will be referred to as consistency criterions of validation processes.

7.2 On the Kolmogorov–Smirnov test

The detection data of an automotive radar sensor are subject to significant stochastic fluctuations and must therefore be taken into account and tolerated to some extent during validation. A statistical approach is evident, especially since this concept is already used for data-based modeling (see chapter 5). The question whether the distributions of simulated and real radar detections match, inevitably leads to statistical tests.
Having both real and synthetic data available, it is always just a sample. Thus, the underlying distributions can only be estimated. Typically, two samples result in two different estimates. Even though they are samplings from one and the same distribution. A statistical test (in this case a two-sample test) initially always supposes that the two samples originate from the same population and attempts to clarify whether there is enough evidence to reject this hypothesis. Mathematically more precise:

> *Suppose simulated and real radar detections originate from the same underlying distribution, is it astonishing – in a statistical sense – to observe the distributions in the two data sets?*

Suitable tests for this purpose are, for example, the two-sample Kolmogorov-Smirnov test. However, there are also several other tests that may be better in certain cases, e.g., the Cramér-von Mises and the Anderson-Darling test [137]. For the concept of a hypothesis test-based validation, it is not decisive which test is used, at first. Therefore, the procedure is explained in case of a Kolmogorov-Smirnov test. In order to provide in-depth understanding of the procedure, a concrete example serves as a starting point.

As a one-dimensional numerical example, the detections from Fig. 7.1 are projected onto the direction of motion, i.e., the longitudinal axis. The origin of the coordinate system is the front of the bounding box. Hence, the one-dimensional projection of real and synthetic detections are given by

$$Z_{\text{real}} = \{0.01, 0.15, 0.5, 1.1, 2.35, 3.7\} \quad \text{and} \quad Z_{\text{sim}} = \{0.05, 0.30, 1.05, 3.35\}\,. \tag{7.7}$$

Important for the purpose of the test are their distribution functions. As a reminder: The distribution function F_X of a random variable X at the point $x \in \mathbb{R}$ reflects the probability for the event $\{X \leqslant x\}$. Therefore one also writes

$$F_X(t) = P\,(X \leqslant t)\,. \tag{7.8}$$

The distribution function of the two samples Z_{real} and Z_{real} are estimated by the empirical distribution function. For any given value $x \in \mathbb{R}$, the empirical distribution function gives the empirical probability of having sample values that are smaller or equal than x. In terms of the identity function $\mathbb{1}$, the empirical distribution functions can be expressed by

$$F_{\text{real}}(x) = \frac{1}{n}\sum_{i=1}^{n} \mathbb{1}\left(z_{\text{real}}^{(i)} \leqslant x\right) \qquad \text{and} \qquad F_{\text{sim}}(x) = \frac{1}{m}\sum_{i=1}^{m} \mathbb{1}\left(z_{\text{sim}}^{(i)} \leqslant x\right). \tag{7.9}$$

The question whether the observations are astonishing in a statistical sense (and thus the null hypothesis has to be rejected) is decided by the maximum difference $d_{n,m}$ of the two empirical distributions. Indeed, due to the strong law of large numbers and the theorem of Gliwenko-Cantelli, the two distribution functions converge uniformly against the actual underlying distribution functions. Consequently, $d_{n,m}$ also estimates the difference between the actual distributions. This maximum difference is given by

$$d_{n,m} = \sup_{x} |F_{\text{real}}(x) - F_{\text{sim}}(x)|\,. \tag{7.10}$$

In case of the numerical example, this implies $d_{n,m} = 0.25$. Whether this value is tolerable or not depends on the level of significance α and the critical value $d_{n,m,\alpha}$, which can be determined from this. For sufficiently large detection numbers ($n, m \geqslant 12$), one can approximately calculate this value according to [138] and [139] by

$$d_{n,m,\alpha} = c_\alpha \sqrt{\frac{n+m}{nm}} \qquad \text{mit} \qquad c_\alpha = \sqrt{\frac{\ln(2) - \ln(\alpha)}{2}}. \tag{7.11}$$

In the previous example, the critical value can be determined using a lookup table [140]. With a significance level $\alpha = 0.05$, $n = 6$ and $m = 4$ one finds $d_{n,m,\alpha} = 0.538$. Therefore, the condition

$$d_{n,m,\alpha} > d_{n,m} \tag{7.12}$$

is satisfied. Thus, there is not enough statistical evidence to reject the hypothesis that both detection lists come from one and the same distribution. Consequently, the model is considered to be valid. As in many cases, however, several questions remain to be answered. In particular, the significance level raises the question how to interpret the

		Reality	
		$F_{\text{real}} = F_{\text{sim}}$	$F_{\text{real}} \neq F_{\text{sim}}$
Validation	"proper model"	correct validation with probability $1-\alpha$	False Negative with probability β
	"no proper model"	False Positive with probability α	correct validation with probability $1-\beta$

Fig. 7.2. Table of error types. When judging a radar sensor model by means of a hypothesis test, two different errors can arise. The test only controls the probability α. The probability β is related to the power of a test. Although an estimate for β is desirable from the simulation point of view, this is of less significance compared to an estimate of α.

results in order to obtain an intuitive validation criterion. As in many cases, however, several questions remain to be answered. In particular, the significance level raises the question how to interpret the results in order to obtain an intuitive validation criterion. As is the case with any classifier, there are two possible types of error. So-called false positive errors (type I errors) arise when real and synthetic data originate from the same distribution, but the data sample deviates significantly. These errors occur with probability α when real and synthetic data are from one and the same distribution and result in an incorrect rejection of the model (see Fig. 7.2).
The other type of error is called false negative (type II error). If real and synthetic data do not originate from one and the same distribution, it may nevertheless happen that the data samples are very similar by chance and the Kolmogorov-Smirnov test consequently comes to the conclusion that it is not legitimate to reject the initial assumption. This error occurs with an unknown, uncontrollable probability β. However, from a simulation point of view, this type of error is tolerable. In fact, one would use an actually invalid model, which nevertheless generates data that seem to agree well with the real detection distributions to some extent. For more details on the two-sample Kolmogorov-Smirnov test see [139, 141] and for a precise mathematical formulation see [142].
The previous example served as an illustration of the basics. In fact, in order to validate the spatial detection distribution of a radar sensor model, a two-dimensional distribution must be considered. Since the Kolmogorov-Smirnov test mainly relies on the empirical distribution functions, it is necessary to clarify how this concept can be transferred to two dimensions, at first. As in the one-dimensional case, the empirical distribution function at the point x indicates the empirical probability of having sample values smaller or equal than x, the canonical generalization of the empirical distribution function results in the definition of

$$F_{\text{real}}(x, y) = \frac{1}{n}\sum_{i=1}^{n} \mathbb{1}\left((x, y) \in Q_{xy}^{(1)}\right) \quad \text{with} \quad Q_{xy}^{(1)} = \left\{(u, v) \in \mathbb{R}^2 \,\|\, u \leqslant x, v \leqslant y\right\}. \tag{7.13}$$

However, instead of the quadrant $Q_{xy}^{(1)}$, any of the other three quadrants could have been chosen. Of course, this would have been possible in the one dimensional case either. In fact, instead of using the definition from equation (7.9), one could have used the opposite definition as well. Since

$$p\left(z_{\text{real}}^{(i)} \geqslant x\right) = 1 - p\left(z_{\text{real}}^{(i)} < x\right) \tag{7.14}$$

holds, however, this is of no consequence for the maximal difference $d_{n,m}$ and thus also not for the test result. For the two-dimensional test, though, this means all the remaining quadrants have to be taken into account as well, i.e.,

$$\begin{aligned} Q_{xy}^{(2)} &= \left\{(u,v) \in \mathbb{R}^2 \,\|\, u \geqslant x, v \leqslant y\right\} \\ Q_{xy}^{(3)} &= \left\{(u,v) \in \mathbb{R}^2 \,\|\, u \leqslant x, v \geqslant y\right\} \\ Q_{xy}^{(4)} &= \left\{(u,v) \in \mathbb{R}^2 \,\|\, u \geqslant x, v \geqslant y\right\}. \end{aligned} \tag{7.15}$$

As a consequence, it is no longer sufficient to calculate the maximal difference $d_{n,m}$ with respect to a single quadrant. Instead, all quadrants must be taken into account, i.e.,

$$d_{n,m} = \sup_{x,y,i} \left| F_{\text{real}}^{(k)}(x,y) - F_{\text{sim}}^{(k)}(x,y) \right|. \tag{7.16}$$

In order to actually calculate the maximal difference, the data must be ordered first, consequently, the computation time of the test increases considerably. The original test introduced by Peacock in [138] shows a complexity of $\mathcal{O}(n^3)$. More recent algorithms, e.g., the one from Fasano and Franceschini introduced in [143] reduce the complexity to $\mathcal{O}(n^2)$. Even more efficient algorithms have been postulated, but this has already been falsified by Lopes at al. [144].

With regard to the example mentioned at the beginning, the two-dimensional data given in equation (7.1) and (7.2) yields a maximum distance of $d_{n,m} = 0.5$. Lets assume for a moment that the value $d_{n,m,\alpha}$ remains the same in two dimensions. In this case, the model would be considered valid at a significance level of 5%. Since $d_{n,m,\alpha} = 0.48$ (in the one-dimensional case) for $\alpha = 0.1$, the model could not be validated at this level of significance. So far, this served as an example and the test itself does not represent a continuous validation map for radar sensor models as specified by equation (7.4).

Unfortunately, the value $d_{n,m,\alpha}$ does not stay the same. Peacock does not provide a formal derivation for the threshold in [138]. Instead, Monte Carlo simulations are performed to justify a formula that is applicable in almost every case and often even too strict (by a factor of 1.1 - 1.5). For a significance level $\alpha \leqslant 0.2$ and sufficiently large data sets $(n, m \geqslant 10)$, an approximate formula similar to the one-dimensional case from equation (7.11) is obtained. Therefore, the formula

$$d_{n,m,\alpha} = \left(\sqrt{\frac{\log(2) - \log(\alpha)}{2}} + 0.5\right)\left(1 - 0.53\left(\frac{nm}{n+m}\right)^{-0.45}\right)\sqrt{\frac{n+m}{nm}} \tag{7.17}$$

is proposed[1]. The threshold itself has not been derived directly in Peacock's paper in this way, but this can be calculated with ease using the four steps given in [138, p. 626].

[1]Additional tests with different distributions have demonstrated that the results are in reasonable agreement with Fig. 7.2.

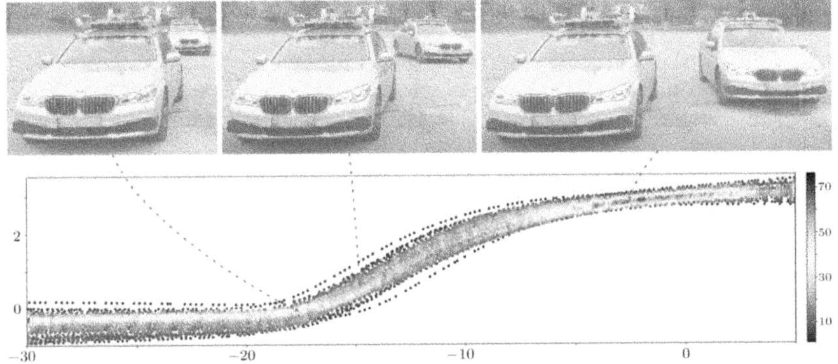

Fig. 7.3. The data basis for testing the validation methodology. During the hundredfold repetition of the overtaking maneuver, the target vehicle moves along an almost identical trajectory. The scatter plot illustrates the target vehicle's positions at the measurement time of the rear left short range radar of the ego vehicle. As one clearly sees, achieving perfect replication is extremely challenging. The heatmap indicates the number of close measurements.

7.3 Applications to radar sensor models

In order to apply statistical hypothesis testing when validating a radar sensor model, certain requirements are mandatory. Furthermore, a validation map for radar sensor models, as specified by equation (7.4), should be defined for the purpose of interpretability. Although the Kolmogorov-Smirnov test can be applied already in case of small detection lists, it is recommended to combine several lists to ensure at least ten detections. Otherwise the approximation given in equation (7.17) cannot be applied. When combining numerous detection lists, an evaluation can be conducted even in case of single detections, as they are typical at certain distances, as well as in case of single detection losses (false negative sensor error). Additionally, multiple detections ensure a more detailed evaluation. But this does explicitly not imply the more the better. Instead, as often is the case, a reasonable compromise needs to be ensured. In principle, there are two conceivable approaches, which are explained in the following and also examined using real radar sensor data (see Fig. 7.3).

7.3.1 Scenario based application

In order to apply the Kolmogorov-Smirnov test, a sufficient amount of radar detections is required. However, if simulated and real measurement times are compared individually, a sufficient quantity is usually not guaranteed. One way to overcome this limitation is to perform the validation for a complete maneuver (scenario). Here, an overtaking maneuver serves as an example and is illustrated in Fig. 7.3. To increase reproducibility, the

Fig. 7.4. Illustration of the measurement point association algorithm. As shown in Fig. 7.3, an exact reproduction of a certain maneuver is nearly unfeasible. Therefore, a reference trajectory must be chosen first (red measuring points) in order to associate the remaining measurements (blue measuring points) according to a nearest neighbor search (depicted in orange). For the subsequent position-based validation, the detection lists can be accumulated.

ego vehicle remains stationary while the target vehicle performs the overtaking maneuver. Thereby, the detection lists of the rear left short range radar sensor are continuously logged (see also Fig. 7.4). The positions of both vehicles are permanently tracked with an additional reference system (differential global positioning system combined with an inertial navigation system), such that the entire maneuver can be re-simulated afterwards. As already mentioned by Rosenberger et al. in [145], using commercial simulation software and further standards like OpenDrive [146] and OpenScenario [147] even an exact re-simulation is nonetheless challenging. For the moment, however, this is neglected, since re-simulation is conducted on single positions without commercial driving simulations.

Finally, the model can be validated with a statistical hypothesis test based on accumulated detection lists. An example of the data sets to be compared in case of a simply parameterized model (cf. chapter 6) is shown in Fig. 7.5. The validation process consistently rejects the model even at a significance level of 5%. If the validation is performed on repeatedly performed maneuvers, the relative frequency of positive validation results can be determined. As a validation criterion, the relative frequency represents a continuous measure as defined by equation (7.4). In case of one hundred overtaking maneuvers, the validity of the simplistic model is 0% at a significance level of 5%.

In order to conclude whether the validation methodology is generally adequate, the question of the consistency defined in section 7.1 must be answered. Only when a sufficient level of consistency has been achieved, the actual test result may be categorized. Without a consistency check, it is also not guaranteed that any model can provide high accuracy data at all. As a reminder, consistency checks involve further validations with additional real data based on nearly identical maneuvers. As before, the validation results are averaged (an analogous consistency check can also be performed on the basis of synthetic data). In case of the exemplary overtaking maneuver, a consistency of 95.7% is obtained. Therefore, if one injects the real sensor data of another maneuver into the simulation, it

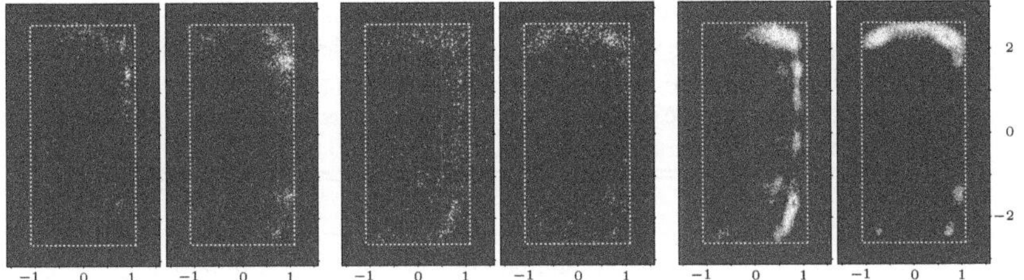

Fig. 7.5. Examples of the detection lists being evaluated. On the left, the comparison of real and simulated detection lists, where the detection lists of 100 maneuvers were accumulated at a certain state of the target vehicle (position based application). In the middle the comparison of the accumulated detection lists of a real and a simulated maneuver (scenario based application). On the right the total accumulation of real and simulated data.

should actually be categorized as a perfect sensor simulation. With 95.7%, the objective validation result comes to almost the same conclusion[2].
The result of the consistency check of more than 95% sounds already very promising. Nevertheless, one may ask why not 100%? In fact, the theoretical optimum is not at 100% percent. If one looks at Fig. 7.2 again, the explanation is straightforward. Since the data sets were recorded with one and the same sensor, the null hypothesis is fulfilled. The statistical hypothesis test with a significance level of α implies that the test classifies the sensor itself as a proper model with a probability of $1 - \alpha$. Thus, in the present example, 95% would have been expected. Consequently, the result can be considered as nearly optimal[3]. And this despite the minor deviations in the trajectories. Finally, it can be concluded that the validation scheme is appropriate to take into account the statistical nature of automotive radar sensor technology and rovides an intuitive and interpretable result beyond simple threshold definitions.

7.3.2 Position based application

An alternative way to apply statistical hypothesis tests to a sufficient detection number is based on a test evaluation at individual time stamps, corresponding to individual measurement points. If one repeatedly performs a certain driving maneuver (as in Fig. 7.3), the spatial detection distribution at certain target vehicle states can be estimated by accumulation of the individual detection lists. For this, however, the measurement times of the individual maneuvers must first be associated. On the one hand, the trajectories

[2]All 100 maneuvers were analyzed regarding trajectory variations prior to the consistency check, resulting in a total of 24 maneuvers cross-validated against each other (276 possible combinations).

[3]This value does not agree with previously introduced results (see [135]), since the one-dimensional threshold $d_{n,m,\alpha}$ was used in that study.

Table 7.1. Summary of validation attempts by means of statistical hypothesis testing. In particular, the scenario based application demonstrates that realistic sensor modeling is within feasible range and hence a promising approach. For the position based application an even stronger focus on accurate replication of driving maneuvers is mandatory in future applications.

	position		scenario	
	real	simulated	real	simulated
real	40.4 %	0 %	95.7 %	0 %
simulated	0 %	15.3 %	0 %	99.6 %

differ slightly and on the other hand, the measurement cannot be triggered and therefore, the target location at the time of measurement deviates due to additional temporal fluctuations.

First, a reference run is selected for the association. The remaining target locations at certain measurement timestamps are mapped to the closest target location of the reference trajectory (see Fig. 7.4). This corresponds to a simple nearest neighbor search. Subsequently, the corresponding detection lists logged at these timestamps are accumulated. An example of a position based accumulation is shown in Fig. 7.5.

The trajectories of all maneuvers are individually emulated during the subsequent re-simulation. Consequently, the synthetic detection lists are also associated and accumulated in an identical manner. For the exemplary overtaking maneuver, a reference trajectory with 102 measurement points is obtained. Hence, the model is finally validated with a statistical hypothesis test at 91 (eleven provide less than ten detections, thus a hypothesis test cannot be performed) target locations each one based on 100 accumulated detection lists. The relative frequency of successfully performed hypothesis tests provides a validation criterion as defined in equation (7.4). The simplistic model is rejected again with a validity of 0% at a significance level of 5%.

As is the case with the scenario based validation method, the question of the consistency of the proposed validation criterion arises once again. For this study, all 102 measurement points are divided into two parts (the first 50 runs and the second 50 runs) and validated against each other. Unfortunately, the data basis is insufficient, so that on the one hand no additional restriction to similar trajectories is possible and on the other hand several measuring timestamps do not contain sufficiently many detections. Therefore, the consistency check results in only 40.4%, even though 95% would have been expected. A consistency check with simulated data (cf. equation 7.6) also yields only 15.3% despite a better data basis. Thus, it is most likely that the trajectories, and thereby the positions of the target vehicle, differ too significantly from each other. In consequence, position-based validation requires excessive focus on an exact reproduction of driving maneuvers. Further tests must be conducted in order to be able to assess the practicability of this validation method.

7.4 Retrospective and future validation challenges

The introduction of statistical hypothesis tests for the evaluation of the spatial distribution of radar detections offers impressive possibilities in the field of sensor model validation. First practical tests based on an overtaking maneuver have shown the general applicability and have also highlighted once more where to focus on in future. Table 7.1 provides a summary of the results.

High variance in real radar data confronts validation with major challenges. However, by cross-validating real data, it has been demonstrated that statistical hypothesis tests can overcome these issues. This led to the definition of consistency criteria for validation procedures of radar sensor models. In this way, it was demonstrated that the development of adequate sensor models (with respect to the presented validation methodology) is within a feasible range. Although it still remains a significant challenge. Due to the limited reproducibility of individual driving maneuvers, the scenario based application is currently to be preferred over a position based applications.

The interpretability and intuitiveness of the validation methodology facilitates a transparent and comprehensible presentation of results even in interdisciplinary environments with numerous stakeholders. Thus, for the first time, objective and purposive criteria for radar sensor model validation have been presented. However, the validation methodology only refers to a small subset of overall model validation. In future tests, it should be examined to what extent other radar characteristics can be validated with this approach and whether a direct application in the sensor coordinate system is equally practicable.

Conclusion and recap of the fourth research question: The proposed validation scheme for the evaluation of radar point cloud models is based on statistical hypothesis testing. It is appropriate to take into account the statistical nature of automotive radar sensor technology and provides an intuitive and interpretable result beyond simple threshold definitions. The presented consistency criterion is achieved within the scope of the significance level. In other words, the validation methodology reliably recognizes real data and rejects simulation data just as faithfully. However, it is necessary to investigate whether and which model data pass the validation and how to choose the significance level. Furthermore, the approach is also suitable for the comparison of different sensor models. For example, an adaptation of the significance level can be used to determine whether a model will pass the validation test sooner than a comparable sensor model and thus should be preferred over the latter.

Chapter 8

Conclusion and prospective challenges

The preceding chapters dealt with the simulation of radar point clouds in standardized simulation environments like the Open Simulation Interface. Since this topic will be of increasing importance for simulative validation of advanced driver assistence systems in the future, the most important results of the previous four chapters will be summarized and presented concisely. Finally, future research topics regarding radar sensor modeling and validation will be discussed. Their practicability and potential will also be elaborated.

8.1 Recap of the radar point cloud simulation

The use of a standardized sensor simulation framework inevitably leads to a significant abstraction regarding the description of the simulated scene. As a result, a detailed physical simulation is usually not feasible. Nevertheless, a radar sensor must be virtualized as detailed as possible in order to simulatively test and validate autonomous driving functions. Since individual software components of the sensor are often proprietary, it is quite common for automotive car manufacturers to operate on the specified output format directly. Due to the lack of technical information, but also on account of available real data, data-based methods are often preferred.

In order to be able to utilize data-based algorithms, though, an extensive amount of real sensor data, including reference sensors, is required. In numerous other disciplines, the lack of data has already led to the use of synthetically generated data for the training of certain algorithms. In principle, this approach is also suitable for radar sensor models. For example, one could use a detailed model that is a priori not suitable for standardized environments to provide training data for data-based approaches. This would considerably reduce the amount of real test drives with multiple vehicles that are equipped with reference sensors, for example. However, the first question that arises is which high-fidelity models are suitable for this purpose. The first research question therefore dealt with the topic of whether the ray-cone-tracing based approach presented in [25] is suitable for this purpose and whether it has the necessary accuracy for generating synthetic training data.

In chapter 4 the methods of ray-cone-tracing were investigated in more detail. An essential feature of this method is the beam expansion due to reflections at curved surfaces. A detailed analysis of the approximation errors has shown which number of initially emitted rays is actually necessary to stay within an acceptable range of detail. In particular, it has been demonstrated that only few multiple reflections can be realized in practice. Due to the analysis and additional modeling complexities, it was concluded that the model cannot be used to generate synthetic training data at the moment and real data comprising reference measurements have to be used for the current time.

Well-known data-based methods have not yet been applied to the simulation of radar point clouds. Therefore, the second research question dealt with the topic of applicability of kernel density estimation methods and deep generative networks in the field of radar sensor modeling at a detection list level. In chapter 5 four modeling approaches were presented. One approach was based on the kernel density estimation method and three others were based on variational autoencoders and generative adversarial networks. Since both, neural networks and kernel density estimation, require a large amount of data, synthetic training data were generated using a two-dimensional ray-tracing method for a general comparison. The evaluation of the learning capacities demonstrated that classical statistical methods are by no means inferior to neural networks and even perform significantly better in certain aspects and should therefore be preferred in case of doubt.

The tremendous demand for data in purely statistical methods raises questions about alternative modeling approaches. Furthermore, in the course of validating automated driving functions, it is often necessary to parameterize models individually for worst-case situations. The third research question was dedicated to this specific topic. In chapter 6, a two-dimensional ray-tracing based radar sensor model was introduced that is capable of being deployed in standardized environments and can be improved throughout all phases of the development process. As a simple generic model, the sensor model can be used in the early stages of development. With real sensor data available in later stages, but to a much lesser extent than in the purely data-based case, the model can be continuously refined to match the real sensor.

On the one hand, parameterizability is preserved and on the other hand, the underlying principles create credibility and confidence for prospective user. Furthermore, the model guarantees reliable data especially if certain ground truth states are not included in the training data. Additional attention was devoted to the modeling of certain micro-Doppler effects. A three-dimensional ray-tracing approach was used to investigate the visually visible velocity profile of rims. Based on an exemplary rim, it could be illustrated how a probability density function for the velocity distribution of a measurement – in case of detections from the wheels – can be derived and how this function may be used in combination with the hybrid modelling approach. The consideration of such effects will be of increasing importance in the future, since this detailed speed information is already used by tracking algorithms today.

Finally, there always remains the question of the actual realism of certain modeling approaches. Especially the validation of point cloud based radar sensor models is very difficult in practice. The fourth and last research question dealt with the definition of practical validation criteria. Chapter 7 introduced the concept of consistency of validation concepts by providing a simplified example. According to this concept, it is important, especially due to the random noise properties of radar detection lists, that validation concepts guarantee the sensor to be a valid model of itself to a certain extent. This idea, as well as commonly implemented statistical sub-modules of radar sensor models, therefore literally demand a validation concept based on statistical hypothesis testing. Two validation methods based on a two-dimensional Komogorov-Smirnoff test have proven that consistency requirements can be satisfied in principle and that real sensor data can be reliably distinguished from synthetic data. However, this requires a reliable and highly precise reproduction of individual driving maneuvers, which involves a certain amount of effort in practice. Another major advantage of the approach is its intuitiveness and interpretability of the validation results in terms of percentages, which is often demanded.

8.2 Lessons learned and future recommendations

The present thesis is dedicated to the topic of detection list based radar sensor modeling in standardized simulation environments such as the Open Simulation Interface, for instance. Although many experts believe ray-tracing based methods will prevail in the long term, standardized simulation architectures remain of great interest due to the popularity across automotive manufacturers. At this point in time, many important topics remain to be investigated as part of further specialized studies. In the following, an assessment of some of these topics is given. Furthermore, several yet unsolved questions are addressed.

Currently, it can be assumed that purely data-based models based on deep generative networks will not be of any importance in practice. Such models are hardly generic, are not parametrizable and do not guarantee any type of convergence. Therefore, it is conceivable that more and more data may even lead to worse results. Furthermore, the required expenditures due to additionally required reference sensors are nearly unmanageable. If so much effort is expended on sensor data collection involving multiple vehicles including reference systems in the course of modeling, the actual objective of saving real test drives may even be missed. Main lesson is that the best results are generally obtained by a judicious trade-off between simplified physical modelling and quantitative data analysis. However, the generation of synthetic training data in the case of models based on kernel density estimation techniques remains a possibility. As mentioned in the thesis, there are still numerous and often similar questions in case of high-fidelity radar models. The generation of realistic training data is therefore not yet in sight.

The hybrid modeling approach presented is most promising for future simulations of radar point clouds. However, the full potential of the approach has not yet been exploited. For example, the question still remains whether the model is capable of achieving a correspondingly high degree of validity with respect to the validation procedure presented and, if so, how to choose an appropriate level of significance.

If the two-dimensional ray-tracing based approach achieves a high degree of validity, this actually only applies to a specific maneuver and a specific vehicle model. In general, the idea is based on the approach of proving validity on few scenarios and consequently establishing model confidence. In contrast to purely data-based models, a simple physical principle guarantees reasonable sensor data in further test scenarios and thus establishes a basic level of model confidence.

Although the validation does not require a large amount of data, the demands on data quality are significant. Even minor deviations during the reproduction of the maneuvers can affect the result. Therefore, an exact reproducibility of individual maneuvers is of great importance for the scenario selection. Moreover, it is important to determine potential limits of tolerance.

Currently, radar sensor modeling generally does not take enough account of vulnerable road users. Separate simulation solutions exist, but no holistic sensor modeling approaches are available. Hence, an promising topic is the future refinement of the hybrid modeling concept to consider vulnerable road users such as motorcyclists, bicyclists, and pedestrians.

Finally, for the presented hybrid modeling approach, the consideration of additional effects is certainly interesting and indeed important. Examples are interference effects of various radar sensors, as well as weather effects. If possible, such effects should be tested in the context of worst-case simulations by means of parameterization, since a purely statistical consideration appears to be difficult to retrace.

Nomenclatur

Radar notation

λ	wave length (also used as parameter (expected value) of the Poisson distribution)
R	distance between radar sensor and target
c_0	speed of light
B	bandwidth of the frequency modulation
T	chirp or sweep time
τ_0	time of flight
n_{C}	number of chirps
n_{Rx}	number of receive antennas
f_{AD}	sampling frequency of the analog-to-digital converter
f_{c}	carrier frequency, i.e., the smallest frequency of a chirp
$\bar{f}_h$	mean frequency within a time interval h
f_{Tx}	transmit frequency of a radar sensor
f_{Rx}	frequency received by the radar sensor after reflection
Δf	difference frequency, i.e., the difference of transmit and receive frequency
$u_{\mathrm{Tx}}, u_{\mathrm{Rx}}$	transmit signal, receive signal
u_{Δ}	intermediate frequency signal
u_*	intermediate frequency signal before low-pass filtering, i.e., $u_* = u_{\mathrm{Tx}}u_{\mathrm{Rx}}$
$\varphi_{\mathrm{Tx}}, \varphi_{\mathrm{Rx}}$	phase of the transmit and receive signal

$\Delta\varphi$ phase difference (also used as angular distance of transmitted rays in the hybrid modeling approach)

$X_{i,j}$ stochastic variable representing the cells of a range-velocity map

μ average noise power in a cell of a range-velocity map

α scaling factor for the detection threshold $\Theta = \alpha\mu$

Θ threshold for a detection in the range-velocity map

p_{FP} probability for a false positive detection

γ Lorentz factor used for the Lorentz transformation (also denotes an angle between the x-axis and the normal vector of the sphere when deriving the caustic distance)

d distance between receive antennas of a radar sensor

d_s auxiliary distance used for the derivation of the angular frequency bins of the third Fourier transform, see Fig. 1.4

Derivation of caustic distance

$\Delta\alpha$ opening angle of the transmitted beam

$\Delta\alpha'$ opening angle of the reflected beam

$\Delta\alpha'_{\mathrm{approx}}$ approximated opening angle of the reflected beam

p length of the ray, i.e., distance between reflection point and transmitter

q caustic distance, i.e., distance between reflection point and the virtual point of transmission of the reflected beam

q_{approx} approximated caustic distance

(x_q, y_q) coordinates of the reflection point on the surface

(x_+, y_+) auxiliary coordinates, see Fig. 4.2

t_α abbreviation used for $t_\alpha = \tan(\theta_0 + \Delta\alpha)$

$\varepsilon_q, \varepsilon_{\Delta\alpha'}$ relative error of the caustic distance and the opening angle when approximation formulas are used

Statistical point cloud simulation

$\hat{f}_n$ kernel density estimation based on n observations(the hat-notation is also used for the approximation of a measurement process and for the discrete Fourier transform of a vector)

M measurement process

$\mathcal{X}$	set of possible ground truth states
$\mathcal{Z}$	set of possible radar point cloud, i.e., detection lists
$X_{\text{sim}}, X_{\text{real}}$	simulated, real ground truth states as a realization of the random variable X
$Z_{\text{sim}}, Z_{\text{real}}$	simulated real radar point clouds as a realization of the random variable Z
$z_{\text{sim}}, z_{\text{real}}$	single simulated, real detections, i.e., elements of a simulated, real radar point cloud, as a realization of realization of the random variable z
$n_{\text{sim}}, n_{\text{real}}$	number of detections within the radar point cloud as a realization of the random variable n
h_0	optimal bandwidth for the kernel density estimation
K_h, K_H	kernel function with bandwidth h or positive definite bandwidth matrix H
ISE	integrated square error of an estimated function
D_{KL}	Kullback-Leibler divergence
$\text{dist}_{\text{BB}}(z)$	distance of a detection z to the closest edge of the bounding box
E_N	quality measure for the evaluation of radar detection models, especially for the number of detections
E_{BB}	quality measure for the evaluation of radar detection models, especially for the shape of a point cloud
E_{KLD}	quality measure for the evaluation of radar detection models based on the Kullback-Leibler divergence
$\hat{Z}$	approximation of a measurement process Z, i.e., a sensor model in terms of a rigorous mathematical formulation (the hat-notation is also used for the kernel density estimation of a function f and for the discrete Fourier transform of a vector)

Hybrid modeling approach

p_{ord}	existence probability (measurement likelihood) of an ordinary hit point
$p_{\text{char}}^{(i)}$	existence probability (measurement likelihood) of the i-th characteristic scattering center
$p_{\text{dist}}^{(i)}$	distance dependent existence probability for detections according to the hybrid model, i indicates a certain characteristic scattering center and $i = 0$ indicates an ordinary hit point
$p_{\text{ang}}^{(i)}$	angular dependent existence probability for detections according to the hybrid model, i indicates a certain characteristic scattering center and $i = 0$ indicates an ordinary hit point

$\mathcal{E}_{\mathrm{ord}}^{(l)}$ measurement deviation distribution (noise distribution) of an ordinary hit point where l indicates the position of (left, right, front, rear)

$\mathcal{E}_{\mathrm{char}}^{(i)}$ measurement deviation distribution (noise distribution) of the i-th characteristic scattering center

ρ density of transmitted rays at distance r, rays per meter arc length at a distance r

ξ state vector describing the ground truth state

r_{w} radius of a wheel

ω_{w} angular velocity of a wheel

v_{r} radial velocity of a point on the bounding box (considering the yaw rate), radial in terms of the sensor coordinate system

v_{rel} relative velocity of a point on the bounding box (considering the yaw rate)

$v_{\mathrm{r,rot}}$ radial velocity of a point on the wheel (considering the yaw rate and rotational motion of the wheel), radial in terms of the sensor coordinate system

$v_{\mathrm{rel,rot}}$ relative velocity of a point on the wheel (considering the yaw rate and rotational motion of the wheel)

$v_{\omega_{\mathrm{w}}}$ velocity vector of a point on the wheel in the coordinate system of the wheel, see Fig. 6.6

$v_{\omega_{\mathrm{w}}}^{\mathrm{w}}$ projection of $v_{\omega_{\mathrm{w}}}$ onto the direction of motion of the wheel in case of no steering angle

$f_{\omega_{\mathrm{w}}}$ radial velocity of all points on an idealized rim (considering the rotational motion of the wheel), radial in terms of the sensor coordinate system

$f_{\mathrm{r,rot}}$ relative velocity distribution of all points on an idealized rim (considering the rotational motion of the wheel and the motion of the vehicle)

Validation of radar sensor models

V validation map in terms of a rigorous mathematical formulation

F_X distribution function of a random variable X

$F_{\mathrm{sim}}, F_{\mathrm{real}}$ empirical (spatial) distribution function of simulated and real radar detections

$d_{n,m}$ maximum difference between F_{sim} and F_{real}

$d_{n,m,\alpha}$ critical value for the Kolmogorov-Smirnov test

$Q_{xy}^{(i)}$	there are four possible quadrants to be considered when calculating $d_{n,m}$ in the two-dimensional case

General mathematical notation

$\mathbb{N}, \mathbb{R}, \mathbb{C}$	set of natural numbers ($0 \notin \mathbb{N}$), set of real numbers, set of complex numbers
$\mathrm{e}, \mathrm{i}, \pi$	Euler's number, imaginary unit, mathematical constant pi
$\lvert x \rvert$	absolute value of x
$\lVert v \rVert$	norm of a vector v
$\overline{z}$	complex conjugation of $z \in \mathbb{C}$, i.e., $\overline{z} = \overline{a + \mathrm{i}b} = a - \mathrm{i}b$
$x \cdot y$	scalar product of two vectors x and y
$H^\top, \det H$	transpose and determinant of the matrix H
$\mathcal{N}(\mu, \sigma^2)$	Gaussian distribution with expectation μ and standard deviation σ
$\lfloor x \rfloor$	floor function, e.g., $\lfloor 5.9 \rfloor = 5$
$\lceil x \rceil$	ceiling function, e.g., $\lceil 1.1 \rceil = 2$
$\sin, \sin^{-1}$	sine function and its inverse
$\cos, \cos^{-1}$	cosine function and its inverse
$\tan, \tan^{-1}$	tangent function and its inverse
$\exp, \log$	exponential function and natural logarithm
$\lim\limits_{n\to\infty} a_n$	limit of a sequence a_n when n tends to infinity
$\sup\limits_{x\in\mathbb{R}} f(x)$	supremum of a function f, e.g., $\sup\limits_{x\in\mathbb{R}} 1 - x^2 = 1$
$\mathcal{F}^{(d)}$	discrete Fourier transform accepting a d-dimensional vector as input
$\hat{z}_k$	Fourier transform of the k-th entry of a vector z (the hat-notation is also used for the kernel density estimation of a function f and for the approximation of a measurement process)

References

[1] KPMG and Center for Automotive Research. *Self-driving cars: The next revolution.* Retrieved February 26th, 2021. www.kpmg.com. 2017.

[2] Daniel Fagnant and Kara Kockelman. "Preparing a Nation for Autonomous Vehicles: Opportunities, Barriers and Policy Recommendations". In: *Transportation Research Part A: Policy and Practice* 77 (2015), pp. 1–24.

[3] Timo Hanke. "Virtual Sensorics: Simulated Environmental Perception for Automated Driving Systems". PhD thesis. Technische Universität München, 2020.

[4] Hermann Winner et al. *Handbuch Fahrerassistenzsysteme.* 3rd ed. Springer Vieweg, 2015.

[5] Markus Maurer, J. Christian Gerdes, and Barbara Lenz. *Autonomes Fahren - Technische, rechtliche und gesellschaftliche Aspekte.* Ed. by Hermann Winner. Springer Open, 2015.

[6] Aptiv et al. *Safety First for Automated Driving.* Online. White Paper. 2019.

[7] Paul Tipler and Gene Mosca. *Physik für Studierende der Naturwissenschaft und Technik.* Ed. by Peter Kersten and Jenny Wagner. 8th ed. Springer Spektrum, 2019.

[8] Steffen Scherr. "FMCW-Radarsignalverarbeitung zur Entfernungsmessung mit hoher Genauigkeit". PhD thesis. Karlsruher Institut für Technologie (KIT), 2016.

[9] Hermann Rohling. "Radar CFAR Thresholding in Clutter and Multiple Target Situations". In: *IEEE Transactions on Aerospace and Electronic Systems* AES-19.4 (1983), pp. 608–621.

[10] Hermann Rohling. "Ordered Statistic CFAR Technique - an Overview". In: *Proc. of the 12th International Radar Symposium (IRS).* 2011, pp. 631–638.

[11] Matthias Kronauge and Hermann Rohling. "Fast Two-Dimensional CFAR Procedure". In: *IEEE Transactions on Aerospace and Electronic Systems* 49.3 (2013).

[12] Mark A. Richards. *Fundamentals of Radar Signal Processing.* McGraw-Hill Education, 2014.

[13] *Pegasus Research Project - Securing Automated Driving Effectively.* Retrieved February 28th, 2021. pegasusprojekt.de.

[14] Alexander Schaermann. "Systematische Bedatung und Bewertung umfelderfassender Sensormodelle". PhD thesis. Technische Universität München, 2020.

[15] Kilian von Neumann-Cosel. "Virtual Test Drive - Simulation umfeldbasierter Fahrzeugfunktionen". PhD thesis. Technische Universität München, 2012.

[16] *Virtual Test Drive - Complete tool-chain for driving simulation applications.* Retrieved February 26th, 2021. mscsoftware.com/product/virtual-test-drive.

[17] *dSPACE ASM - Sensor Simulation.* Retrieved February 26th, 2021. dspace.com.

[18] *IPG CarMaker.* Retrieved February 26th, 2021. ipg-automotive.com.

[19] Timo Hanke et al. *Open Simulation Interface: A generic interface for the environment perception of automated driving functions in virtual scenarios.* Retrieved February 26th, 2021. ei.tum.de/hot. 2017.

[20] *ASAM OSI.* Retrieved February 12th, 2021. asam.net/standards/.

[21] Karl Granström, Marcus Baum, and Stephan Reuter. "Extended Object Tracking: Introduction, Overview and Applications". In: *Journal of Advances in Information Fusion* 12.2 (2017), pp. 139–174.

[22] Martin Holder et al. "Measurements revealing Challenges in Radar Sensor Modeling for Virtual Validation of Autonomous Driving". In: *Proc. of the 21st International Conference on Intelligent Transportation Systems (ITSC).* 2018, pp. 2616–2622.

[23] Tim A. Wheeler et al. "Deep Stochastic Radar Models". In: *Proc. of the Intelligent Vehicles Symposium.* 2017, pp. 47–53.

[24] Nils Hirsenkorn et al. "Virtual sensor models for real-time applications". In: *Advances in Radio Science* (2016), pp. 31–37.

[25] Nils Hirsenkorn et al. "A Ray Launching Approach for Modeling an FMCW Radar System". In: *Proc. of the 18th International Radar Symposium.* 2017, pp. 1–10.

[26] Christina Knill, Alexander Scheel, and Klaus Dietmayer. "A Direct Scattering Model for Tracking Vehicles with High-Resolution Radars". In: *Proc. of the Intelligent Vehicles Symposium.* 2016, pp. 298–303.

[27] Philipp Berthold et al. "An Abstracted Radar Measurement Model for Extended Object Tracking". In: *Proc. of the 21st International Conference on Intelligent Transportation Systems.* 2018, pp. 3866–3872.

[28] Ralph Rasshofer, Johann Rank, and Guangyu Zhang. "Generalized Modeling of Radar Sensors for Next-Generation Virtual Driver Assistance Function Prototyping". In: *Proc. of the 12th World Congress on Intelligent Transport Systems.* Vol. 4. 2005, pp. 2511–2522.

[29] Timo Hanke et al. "Generic Architecture for Simulation of ADAS sensors". In: *Proc. of the 16th International Radar Symposium.* 2015, pp. 125–130.

[30] Timo Hanke et al. "Classification of Sensor Errors for the Statistical Simulation of Environmental Perception in Automated Driving Systems". In: *Proc. of the 19th International Conference on Intelligent Transportation Systems*. 2016, pp. 643–648.

[31] Timo Hanke et al. "Generation and Validation of Virtual Point Cloud Data for Automated Driving Systems". In: *Proc. of the 20th International Conference on Intelligent Transportation Systems*. 2017, pp. 1–6.

[32] Stefan Bernsteiner et al. "Radarsensormodell für den virtuellen Entwicklungsprozess". In: *ATZelektronik* 10.2 (2015), pp. 72–79.

[33] Nils Hirsenkorn. "Modellbildung und Simulation der Fahrzeugumfeldsensorik". PhD thesis. Technische Universität München, 2018.

[34] Diederik P. Kingma and Max Welling. "Auto-Encoding Variational Bayes". In: *Proc. of the International Conference on Learning Representations*. 2014, pp. 1–14.

[35] Ian J. Goodfellow et al. "Generative Adversarial Nets". In: *In Proc. of the 27th International Conference on Advances in Neural Information Processing Systems 27*. 2014, pp. 2672–2680.

[36] Robin Schubert, Norman Mattern, and Roy Bours. "Simulation von Sensorfehlern zur Evaluierung von Fahrerassistenzsystemen". In: *Fahrerassistenzsysteme und Effiziente Antriebe*. Springer Vieweg, 2015, pp. 69–73.

[37] Alexander Schaermann and Timo Hanke. "Generation and Validation of Sensor Models for Automated Driving Systems Using VIRES VTD". In: *Engineering Reality Magazin*. Vol. 9. MSC Software, 2019, pp. 14–17.

[38] Martin Herrmann and Helmut Schön. "Efficient Sensor Development Using Raw Signal Interfaces". In: *Fahrerassistenzsysteme 2018*. Springer, 2019, pp. 30–39.

[39] Carlo van Driesten and Thomas Schaller. "Overall Approach to Standardize AD Sensor Interfaces: Simulation and Real Vehicle". In: *Fahrerassistenzsysteme 2018*. Ed. by Thorsten Bertram. Springer, 2019, pp. 47–55.

[40] R. G. Kouyoumjian, L. Peters, and D. T. Thomas. "A Modified Geometrical Optics Method for Scattering by Dielectric Bodies". In: *IEEE Transactions on Antennas and Propagation* 11.6 (1963), pp. 690–703.

[41] Morris Kline. "An Asymptotic Solution of Maxwell's Equations". In: *Communications on Pure and Applied Mathematics* 4.2-3 (1951), pp. 225–262.

[42] Joseph B. Keller. "Rays, Waves and Asymptotics". In: *Bulletin of the American Mathematical Society* 84.5 (1978), pp. 727–750.

[43] Zhengqing Yun and Magdy F. Iskander. "Ray Tracing for Radio Propagation Modeling: Principles and Applications". In: *IEEE Access* 3 (2015), pp. 1089–1100.

[44] Max Born and Emil Wolf. *Principles of Optics: Electromagnetic Theory of Propagation, Interference and Diffraction of Light*. 7th (expanded) edition. Cambridge University Press, 1999.

[45] Magdy F. Iskander and Zhengqing Yun. "Propagation Prediction Models for Wireless Communication Systems". In: *IEEE Transactions on Microwave Theory and Techniques* 50.3 (2002), pp. 662–673.

[46] Karin Schuler, Denis Becker, and Werner Wiesbeck. "Extraction of Virtual Scattering Centers of Vehicles by Ray-Tracing Simulations". In: *IEEE Transactions on Antennas and Propagation* 56.11 (2008), pp. 3543–3551.

[47] Jürgen Maurer. "Strahlenoptisches Kanalmodell für die Fahrzeug-Fahrzeug-Funkkommunikation". PhD thesis. Universität Karlsruhe (TH), 2005.

[48] Herman Buddendick, Thomas Eibert, and Jürgen Hasch. "Bistatic Scattering Center Models for the Simulation of Wave Propagation in Automotive Radar Systems". In: *Proc. of the German Microwave Conference (GeMiC)*. 2010, pp. 288–291.

[49] Steven G. Parker et al. "OptiX: A General Purpose Ray Tracing Engine". In: *ACM Transactions on Graphics* 29.4 (2010), pp. 1–13.

[50] Alexander Schunert et al. "Simulation Assisted High-Resolution PSI Analysis". In: *International Archives of the Photogrammetry, Remote Sensing and Spatial Information Sciences* 38.7B (2010), pp. 498–503.

[51] Benny Bürger et al. "Real-Time GPU-Based Ultrasound Simulation Using Deformable Mesh Models". In: *IEEE Transactions on Medical Imaging* 32.3 (2013), pp. 609–618.

[52] Karol Majek and Janusz Bedkowski. "Range Sensors Simulation Using GPU Ray Tracing". In: *Proc. of the 9th International Conference on Computer Recognition Systems*. 2016.

[53] Martin Holder et al. "The Fourier Tracing Approach for Modeling Automotive Radar Sensors". In: *Proc. of the 20th International Radar Symposium*. 2019, pp. 1–8.

[54] Bernhard Schick et al. "Sophisticated Sensor Model Framework Providing Realistic Radar Sensor Behavior in Virtual Environments". In: *8. Fachtagung Fahrerassistenz*. 2017.

[55] Vera Kurz et al. "Retroreflective mmWave Measurements to Determine Road Surface Characteristics". In: *Proc. of the Kleinheubach Conference*. 2019, pp. 1–4.

[56] Stefan O. Wald and Frank Weinmann. "Ray Tracing for Range-Doppler Simulation of 77 GHz Automotive Scenarios". In: *Proc. of the 13th European Conference on Antennas and Propagation*. 2019, pp. 1–4.

[57] Sinan Hasirlioglu et al. "Modeling and Simulation of Rain for the Test of Automotive Sensor Systems". In: *Proc. of the Intelligent Vehicles Symposium*. 2016, pp. 286–291.

[58] Sinan Hasirlioglu et al. "Effects of Exhaust Gases on Laser Scanner Data Quality at Low Ambient Temperatures". In: *Proc. of the Intelligent Vehicles Symposium*. 2017, pp. 1708–1713.

[59] Alexander Prinz et al. "Validation Strategy for Radar-Based Assistance Systems under the Influence of Interference". In: *Proc. of the German Microwave Conference (GeMiC)*. 2020, pp. 252–255.

[60] Alexander Prinz et al. "Automotive Radar Signal and Interference Simulation for Testing Autonomous Driving". In: *Proc. of the 4th International Conference on Intelligent Transport Systems*. 2020, pp. 1–17.

[61] Markus Bühren. "Simulation und Verarbeitung von Radarziellisten im Automobil". PhD thesis. Universität Stuttgart, 2008.

[62] Markus Bühren and Bin Yang. "Simulation of Automotive Radar Target Lists using a Novel Approach of Object Representation". In: *Proc. of the Intelligent Vehicles Symposium*. 2006, pp. 314–319.

[63] Markus Bühren and Bin Yang. "Automotive Radar Target List Simulation based on Reflection Center Representation of Objects". In: *Proc. of the International Workshop on Intelligent Transportation (WIT)*. 2006, pp. 161–166.

[64] Merrill I. Skolnik. *Radar Handbook*. McGraw-Hill Education, 2008.

[65] Jürgen Detlefsen. *Radartechnik*. Ed. by Hans Marko. Vol. 18. Nachrichtentechnik. Springer, 1989.

[66] Markus Bühren and Bin Yang. "Extension of Automotive Radar Target List Simulation to consider further Physical Aspects". In: *Proc. of the 7th International Conference on ITS Telecommunications*. 2007, pp. 1–6.

[67] J. Gunnarsson et al. "Tracking vehicles using radar detections". In: *Proc. of the Intelligent Vehicles Symposium*. 2007, pp. 296–302.

[68] Lars Hammarstrand et al. "Extended Object Tracking using a Radar Resolution Model". In: *IEEE Transactions on Aerospace and Electronic Systems* 48.3 (2012), pp. 2371–2386.

[69] Alexander Scheel et al. "Multi-Sensor Multi-Object Tracking of Vehicles Using High-Resolution Radars". In: *Proc. of the Intelligent Vehicles Symposium*. 2016, pp. 558–565.

[70] Philipp Berthold et al. "Radar Reflection Characteristics of Vehicles for Contour and Feature Estimation". In: *Proc. of the Workshop on Sensor Data Fusion: Trends, Solutions, Applications (SDF)*. 2017, pp. 1–6.

[71] Philipp Berthold et al. "A Radar Measurement Model for Extended Object Tracking in Dynamic Scenarios". In: *Proc. of the Intelligent Vehicles Symposium*. 2019, pp. 770–776.

[72] Dominik Kellner. "Verfahren zur Bestimmung von Objekt- und Eigenbewegung auf Basis der Dopplerinformation hochauflösender Radarsensoren". PhD thesis. Universität Ulm, 2016.

[73] Dominik Kellner et al. "Tracking of Extended Objects with High-Resolution Doppler Radar". In: *IEEE Transactions on Intelligent Transportation Systems* 17.5 (2016), pp. 1341–1353.

[74] Kevin Gilholm and David Salmond. "Spatial distribution model for tracking extended objects". In: *IEE Proceedings - Radar, Sonar and Navigation* 152.5 (2005), pp. 364–371.

[75] David J. Sheskin. *Handbook of Parametric and Nonparametric Statistical Procedures.* Third Edition. Chapman & Hall/CRC, 2003.

[76] Nils Hirsenkorn et al. "A Non-Parametric Approach for Modeling Sensor Behavior". In: *In Proc. of the 16th International Radar Symposium.* 2015, pp. 131–136.

[77] Kevin Gilholm et al. "Poisson models for extended target and group tracking". In: *Proc. of Signal and Data Processing of Small Targets.* Vol. 5913. International Society for Optics and Photonics. SPIE, 2005, pp. 230–241.

[78] Peter Broßeit and Bharanidhar Duraisamy. "The Volcanormal Density for Radar-Based Extended Target Tracking". In: *Proc. of the 20th International Conference on Intelligent Transportation Systems.* 2017, pp. 1–6.

[79] Alexander Scheel and Klaus Dietmayer. "Tracking Multiple Vehicles Using a Variational Radar Model". In: *IEEE Transactions on Intelligent Transportation Systems* 20.10 (2019), pp. 3721–3736.

[80] Jasmin Ebert et al. "Deep Radar Sensor Models for Accurate and Robust Object Tracking". In: *Proc. of the 23rd International Conference on Intelligent Transportation Systems.* 2020, pp. 1–6.

[81] Kihyuk Sohn, Honglak Lee, and Xinchen Yan. "Learning Structured Output Representation using Deep Conditional Generative Models". In: *Advances in Neural Information Processing Systems.* 2015, pp. 3483–3491.

[82] Mehdi Mirza and Simon Osindero. "Conditional Generative Adversarial Nets". In: *Computing Research Repository (CoRR) on arXiv.org* (2014).

[83] *ENABLE-S3 - European Initiative to Enable Validation for Highly Automated Safe and Secure Systems.* Retrieved February 28th, 2021. enable-s3.eu.

[84] *SET Level - Simulationsbasiertes Entwickeln und Testen von automatisiertem Fahren.* Retrieved February 28th, 2021. setlevel.de.

[85] William L. Oberkampf and Christopher J. Roy. *Verification and Validation in Scientific Computing.* Cambridge University Press, 2010.

[86] Robert G. Sargent. "Verification and Validation of Simulation Models". In: *Proc. of the Winter Simulation Conference.* 2010, pp. 166–183.

[87] William L. Oberkampf and Timothy G. Trucano. "Verification and Validation Benchmarks". In: *Nuclear Engineering and Design* 238.3 (2008), pp. 716–743.

[88] Erwin Roth et al. "Analysis and Validation of Perception Sensor Models in an Integrated Vehicle and Environment Simulation". In: *Proc. of the 22nd Enhanced Safety of Vehicles Conference.* 2011, pp. 1–9.

[89] Alexander Schaermann et al. "Validation of Vehicle Environment Sensor Models". In: *Proc. of the Intelligent Vehicles Symposium.* 2017, pp. 405–411.

[90] Michael Viehof. "Objektive Qualitätsbewertung von Fahrdynamiksimulationen durch statistische Validierung". PhD thesis. Technische Universität Darmstadt, 2018.

[91] Martin F. Holder et al. "How to evaluate synthetic radar data? Lessons learned from finding driveable space in virtual environments". In: *Proc. of the 13. Workshop Fahrerassistenzsysteme und automatisiertes Fahren.* 2020.

[92] Michael Viehof and Hermann Winner. *Stand der Technik und der Wissenschaft: Modellvalidierung im Anwendungsbereich der Fahrdynamiksimulation.* TUprints - Publikationsservice der Technischen Universität Darmstadt. July 2017.

[93] Osman Balci. "Principles and Techniques of Simulation Validation, Verification, and Testing". In: *Proc. of the 27th Conference on Winter Simulation.* 1995, pp. 147–154.

[94] Jürgen Dickmann et al. "Radar contribution to highly automated driving". In: *Proc. of the 44th European Microwave Conference.* 2014, pp. 1715–1718.

[95] Nils Hirsenkorn et al. "Learning Sensor Models for Virtual Test and Development". In: *Proc. of the 11. Workshop Fahrerassistenz und automatisiertes Fahren.* 2017, pp. 115–124.

[96] Peng Cao, Walther Wachenfeld, and Hermann Winner. "Perception sensor modeling for virtual validation of automated driving". In: *it - Information Technology* 57.4 (2015), pp. 243–251.

[97] Peng Cao. "Modeling Active Perception Sensors for Real-Time Virtual Validation of Automated Driving Systems". PhD thesis. Technische Universität Darmstadt, 2017.

[98] Apostolia Tsirikoglou et al. "Procedural Modeling and Physically Based Rendering for Synthetic Data Generation in Automotive Applications". In: *Computing Research Repository (CoRR) on arXiv.org* (2017).

[99] German Ros et al. "The SYNTHIA Dataset: A Large Collection of Synthetic Images for Semantic Segmentation of Urban Scenes". In: *Proc. of the Conference on Computer Vision and Pattern Recognition (CVPR).* 2016, pp. 3234–3243.

[100] Stephan R. Richter et al. "Playing for Data: Ground Truth from Computer Games". In: *Proc. of the European Conference on Computer Vision (ECCV).* 2016, pp. 102–118.

[101] Leno S. Pedrotti. *Fundamentals of Photonics - Module 1.3 Basic Geometrical Optics.* Ed. by Chandrasekhar Roychoudhuri. SPIE Digital Library, 2008.

[102] Constantine A. Balanis. *Advanced Engineering Electromagnetics.* 2nd ed. John Wiley & Sons, Inc., 2012.

[103] Georges A. Deschamps. "Ray Techniques in Electromagnetics". In: *Proc. of the IEEE* (1972).

[104] Robert G. Kouyoumjian and Prabhakar H. Pathak. "A Uniform Geometrical Theory of Diffraction for an Edge in a Perfectly Conducting Surface". In: *Proceedings of the IEEE* 62.11 (1974), pp. 1448–1461.

[105] Charles H. Holbrow et al. *Modern Introductory Physics*. Springer New York, 2010.

[106] Thomas Eder et al. "Data Driven Radar Detection Models: A Comparison of Artificial Neural Networks and Non Parametric Density Estimators on Synthetically Generated Radar Data". In: *Proc. of the Kleinheubach Conference*. 2019, pp. 1–4.

[107] Christopher Bishop. *Pattern Recognition and Machine Learning*. Springer, 2007.

[108] Emanuel Parzen. "On Estimation of a Probability Density Function and Mode". In: *Annals of Mathematical Statistics* 33.3 (1962), pp. 1065–1076.

[109] Murray Rosenblatt. "Remarks on some Nonparametric Estimates of a Density Function". In: *Annals of Mathematical Statistics* 27.3 (1956), pp. 832–837.

[110] Qi Li and Jeffrey Scott Racine. *Nonparametric Econometrics: Theory and Practice*. Princeton University Press, 2007.

[111] Elizbar A. Nadaraya. "On Non-Parametric Estimates of Density Functions and Regression Curves". In: *Theory of Probability & Its Applications* 10.1 (1965), pp. 186–190.

[112] Mats Rudemo. "Empirical Choice of Histograms and Kernel Density Estimators". In: *Scandinavian Journal of Statistics* 9.2 (1982), pp. 65–78.

[113] Charles J. Stone. "An Asymptotically Optimal Window Selection Rule for Kernel Density Estimates". In: *The Annals of Statistics* 12.4 (1984), pp. 1285–1297.

[114] Jeffrey S. Simonoff. *Smoothing Methods in Statistics*. Springer, 1996.

[115] Frank Rosenblatt. *The Perceptron, a Perceiving and Recognizing Automaton (Project Para)*. Cornell Aeronautical Laboratory, 1957.

[116] Alexey Grigorevich Ivakhnenko. "Polynomial Theory of Complex Systems". In: *IEEE transactions on Systems, Man, and Cybernetics* (1971), pp. 364–378.

[117] Yann LeCun et al. "Backpropagation Applied to Handwritten Zip Code Recognition". In: *Neural Computation* 1.4 (1989), pp. 541–551.

[118] Alex Krizhevsky, Ilya Sutskever, and Geoffrey E. Hinton. "ImageNet Classification with Deep Convolutional Neural Networks". In: *Advances in Neural Information Processing Systems* 25 (2012), pp. 1097–1105.

[119] *Christie's Auction: Edmond de Belamy (Generative Adversarial Network print)*. Retrieved February 22th, 2021. christies.com.

[120] Tero Karras, Samuli Laine, and Timo Aila. "A Style-Based Generator Architecture for Generative Adversarial Networks". In: *Proceedings of the IEEE/CVF Conference on Computer Vision and Pattern Recognition*. 2019, pp. 4401–4410.

[121] Seung Wook Kim et al. "Learning to Simulate Dynamic Environments with GameGAN". In: *Proc. of the IEEE/CVF Conference on Computer Vision and Pattern Recognition*. 2020, pp. 1231–1240.

[122] Ian Goodfellow, Yoshua Bengio, and Aaron Courville. *Deep Learning*. deeplearningbook.org. MIT Press, 2016.

[123] Lars Mescheder, Andreas Geiger, and Sebastian Nowozin. "Which Training Methods for GANs do actually Converge?" In: *Proc. of the International Conference on Machine Learning.* 2018, pp. 3481–3490.

[124] Martin Arjovsky and Léon Bottou. "Towards Principled Methods for Training Generative Adversarial Networks". In: *Proc. of the 5th International Conference on Learning Representations.* 2017, pp. 1–17.

[125] Tim Salimans et al. "Improved Techniques for Training GANs". In: *Proc. of the 30th International Conference on Neural Information Processing Systems.* 2016, pp. 2234–2242.

[126] Anders Boesen Lindbo Larsen et al. "Autoencoding beyond pixels using a learned similarity metric". In: *Proc. of the Intern. Conference on Machine Learning Research.* Ed. by Maria Florina Balcan and Kilian Q. Weinberger. Vol. 48. 2016, pp. 1558–1566.

[127] Jianmin Bao et al. "CVAE-GAN: Fine-Grained Image Generation through Asymmetric Training". In: *Proc. of the International Conference on Computer Vision (ICCV).* 2017, pp. 2764–2773.

[128] Kurt Hornik. "Approximation Capabilities of Multilayer Feedforward Neworks". In: *Neural Networks* 4.2 (1991), pp. 251–257.

[129] John C. Stover. *Optical Scattering: Measurement and Analysis.* 2nd ed. SPIE Optical Engineering Press, 1995.

[130] *Continental Automotive SRR520 (Specifications).* Retrieved January 27th, 2021. continental-automotive.com.

[131] Dominik Kellner et al. "Wheel Extraction based on Micro Doppler Distribution using High-Resolution Radar". In: *Proc. of the International Conference on Microwaves for Intelligent Mobility.* 2015, pp. 1–4.

[132] *Open Simulation Interface (GitHub).* Retrieved January 28th, 2021. github.com.

[133] Paul Riekert and Theo-Ernst Schunck. "Zur Fahrmechanik des gummibereiften Kraftfahrzeugs". In: *Ingenieur-Archiv* 11.3 (1940), pp. 210–224.

[134] Rodrigo Pérez et al. "Single-Frame Vulnerable Road Users Classification with a 77 GHz FMCW Radar Sensor and a Convolutional Neural Network". In: *Proc. of the 19th International Radar Symposium.* 2018, pp. 1–10.

[135] Thomas Eder et al. "Szenarienbasierte Validierung eines hybriden Radarmodells für Test und Absicherung automatisierter Fahrfunktionen". In: *Automobil-Sensorik 3.* Ed. by Thomas Tille. Springer, 2020, pp. 21–43.

[136] *CST STUDIO SUITE - Asymptotic solver for electromagnetic simulation.* Retrieved February 28th, 2021. www.3ds.com.

[137] Nornadiah Mohd Razali and Yap Bee Wah. "Power Comparisons of Shapiro-Wilk, Kolmogorov-Smirnov, Lilliefors and Anderson-Darling Tests". In: *Journal of Statistical Modeling and Analytics* 2.1 (2011), pp. 21–33.

[138] John A. Peacock. "Two-dimensional goodness-of-fit testing in astronomy". In: *Monthly Notices of the Royal Astronomical Society* 202.3 (1983), pp. 615–627.

[139] Donald Knuth. *Art of Computer Programming, Volume 2 (Seminumerical Algorithms)*. 3rd. Vol. 2. Addison-Wesley Professional, 1997.

[140] James F. Rohlf and Robert R. Sokal. *Statistical Tables*. 3rd Edition. W. H. Freeman and Company, 1995.

[141] Jürgen Hedderich and Lothar Sachs. *Angewandte Statistik - Methodensammlung mit R*. Springer Spektrum, 2018.

[142] John W. Pratt and Jean D. Gibbons. *Concepts of Nonparametric Theory*. Springer New York, 1981.

[143] G. Fasano and A. Franceschini. "A multidimensional version of the Kolmogorov-Smirnov test". In: *Monthly Notices of the Royal Astronomical Society* 225 (1987), pp. 155–170.

[144] Raul H.C. Lopes, Ivan Reid, and Peter R. Hobson. "The two-dimensional Kolmogorov-Smirnov test". In: *Proc. of the International Workshop on Advanced Computing and Analysis Techniques in Physics Research*. 2007.

[145] Philipp Rosenberger et al. "Towards a Generally Accepted Validation Methodology for Sensor Models - Challenges, Metrics, and First Results". In: *Proc. of the Graz Symposium Virtual Vehicle (GSVF)*. 2019, pp. 1–13.

[146] *ASAM OpenDRIVE®*. Retrieved February 12th, 2021. asam.net/standards/.

[147] *ASAM OpenSCENARIO®*. Retrieved February 12th, 2021. asam.net/standards/.

Index

A

acceptance . . . 77
acceptance-rejection method . . . 45
adaptive cruise control . . . 2
advanced driver assistance systems . . . 2
amplitude . . . 5
analog-to-digital converter . . . 10
Anderson-Darling test . . . 79
angle estimation . . . 9
angular density . . . 61
approximation error . . . 37
asymmetric distribution . . . 63
autoencoder . . . 52

B

backpropagation . . . 51, 55
backscattering center . . . 64
bandwidth . . . 4
Bayesian inference . . . 53
beam expansion . . . 33
binary classifier . . . 54
Black-Box-Model . . . 17
bounding box . . . 15
bump function . . . 62

C

CA-CFAR . . . 13
carrier frequency . . . 4
caustic distance . . . 33
cell-averaging procedure . . . 13
CFAR algorithm . . . 12
characteristic scattering center . . . 64, 74
chirp . . . 4
chirp time . . . 4
computer vision . . . 51
conditional GAN . . . 25
conditional probability . . . 45
conditional VAE . . . 25
consistency criterion . . . 79
constant false alarm rate . . . 10, 12
convolutional neural network . . . 51
cosine window . . . 62
Cramér-von Mises test . . . 79
critical value . . . 80
cross-validation . . . 48
Crystal Ball function . . . 63
curve-fitting . . . 74
curved surface . . . 33

D

data-based optimization . . . 73
data-based sensor model . . . 17
deep generative networks . . . 17
deep learning . . . 51
detection list . . . 13
deterministic sensor model . . . 18
difference frequency . . . 4
differentiable loss . . . 56

differential global positioning system 44, 73, 74, 84
discriminator . . . 54
distance-dependency . . . 73
Doppler effect . . . 6

E

emergency break . . . 2
empirical distribution function . . . 80
Epanechnikov kernel . . . 47
error model . . . 18
existence probability . . . 60, 73
expectation value . . . 3
expectation–maximization algorithm . 74
extended-object-tracking-algorithms . . 17

F

false positive probability . . . 13
feature matching . . . 56
FMCW radar . . . 4
Fourier transformation . . . 10
frequency
  instantaneous . . . 5
  intermediate . . . 5
frequency bins . . . 11
frequency modulated continuous wave . 4

G

GAN . . . 25
Gaussian . . . 47
Gaussian mixture model . . . 25
generative adversarial network . . . 25
generative adversarial networks . . . 54
generative neural networks . . . 51
generator . . . 54
grid search algorithm . . . 49
ground truth . . . 15
ground truth sensor model . . . 17
guard cells . . . 13

H

headlights . . . 64
hybrid approach . . . 59

I

idealized model . . . 19
image analysis . . . 51
inertial navigation system . 44, 73, 74, 84
integrated square error . . . 48
interpretability . . . 77, 83
intuitiveness . . . 77

K

kernel density estimation . . . 46
kernel density estimator . . . 17
Kolmogorov-Smirnov test . . . 79
Kullback-Leibler divergence . . . 53

L

label smoothing . . . 56
latent space . . . 52
law of large numbers . . . 80
learning capacities . . . 56
leave-one-out . . . 49
level of significance . . . 80
log-normal distribution . . . 63
Lorentz factor . . . 8
Lorentz transformation . . . 8

M

machine learning . . . 51
machine translation . . . 51
mapping . . . 44
maximum likelihood estimation . . . 54
measurement likelihood . . . 60, 73
measurement model . . . 18
measuring process . . . 44
micro-Doppler effect . . . 24, 64
mirror formula . . . 34
modeling levels . . . 17
multipath propagation . . . 21, 38, 75

N

Nash-equilibrium . . . 54
nearest neighbor search . . . 86
neural network . . . 51
non-parametric sensor model . . . 25

O

occlusion . . . 75
occupancy grid map . . . 27, 78
opening angle . . . 35
ordered statistics . . . 14
OS-CFAR . . . 14

OSCA-CFAR 14
OSI - Open Simulation Interface 15
over-smoothing 48

P

parameterization 73
parametric sensor model 25
perceptron 51
phase 5
phenomenological sensor model 17
physical model 18
point cloud 13
point of impact 35
Poisson distribution 2
principal component analysis 52
pulse radar 4

R

radar cube 21
radar point cloud 4
random variable 45
range-Doppler map 12
range-velocity map 12
range-velocity-angle map 21
raw data model 18
ray cone tracing 33
ray density 73
ray-tracing based sensor model 17
Rayleight roughness criterion 61
re-parameterization 52
re-parametrization trick 53
receive frequency 4
reconstruction loss 54
reference trajectory 86
relative error 37
relative velocity 63
reproducibility 83
rim 64

S

SAE Level 2
sawtooth frequency modulation 4
scenario based validation 85
sensor data interface 15
signal processing 4
single-track model 66
skew-normal distribution 63
small-angle approximation 34
speech recognition 51
statistical model 18
statistical tests 79
steering angle 66, 68
stochastic gradient descent 55
stochastic random variable 13
sufficiently regular 52
sweep time 4

T

tail lights 64
target detection procedure 13
test kilometers 2
threshold-based validation 77
time of flight 4
tire 64
tophat kernel 47
transmit frequency 4
two-sample test 79
type I error 81
type II error 81

U

under-smoothing 48
underbody clearance 65

V

VAE 25, 51
VAE-GAN 56
validation 25, 77
validation strategy 3
variational autoencoder 25, 51
variational distribution 53
velocity distribution 70
velocity measurement 71
verification 25
visible region 71

W

well-organized 52
wheel housing 64
wheel speed effects 64
window function 62

Y

yaw angle 68
yaw rate 63

www.ingramcontent.com/pod-product-compliance
Ingram Content Group UK Ltd.
Pitfield, Milton Keynes, MK11 3LW, UK
UKHW021653190726
13853UKWH00001B/234